Google Cloud Platform for Data Science

A Crash Course on Big Data, Machine Learning, and Data Analytics Services

Dr. Shitalkumar R. Sukhdeve
Sandika S. Sukhdeve

apress®

Google Cloud Platform for Data Science: A Crash Course on Big Data,
Machine Learning, and Data Analytics Services

Dr. Shitalkumar R. Sukhdeve
Gondia, India

Sandika S. Sukhdeve
Gondia, Maharashtra, India

ISBN-13 (pbk): 978-1-4842-9687-5
https://doi.org/10.1007/978-1-4842-9688-2

ISBN-13 (electronic): 978-1-4842-9688-2

Managing Director, Apress Media LLC: Welmoed Spahr
Acquisitions Editor: Susan McDermott
Development Editor: Laura Berendson
Coordinating Editor: Jessica Vakili

Cover designed by eStudioCalamar

Cover image from www.freepik.com

Distributed to the book trade worldwide by Apress Media, LLC, 1 New York Plaza, New York, NY 10004, U.S.A. Phone 1-800-SPRINGER, fax (201) 348-4505, e-mail orders-ny@springer-sbm.com, or visit www.springeronline.com. Apress Media, LLC is a California LLC and the sole member (owner) is Springer Science + Business Media Finance Inc (SSBM Finance Inc). SSBM Finance Inc is a **Delaware** corporation.

For information on translations, please e-mail booktranslations@springernature.com; for reprint, paperback, or audio rights, please e-mail bookpermissions@springernature.com.

Apress titles may be purchased in bulk for academic, corporate, or promotional use. eBook versions and licenses are also available for most titles. For more information, reference our Print and eBook Bulk Sales web page at http://www.apress.com/bulk-sales.

Any source code or other supplementary material referenced by the author in this book is available to readers on GitHub (https://github.com/Apress). For more detailed information, please visit https://www.apress.com/gp/services/source-code.

Paper in this product is recyclable

To our parents, family, and friends.

Table of Contents

About the Authors

Dr. Shitalkumar R. Sukhdeve is an experienced senior data scientist with a strong track record of developing and deploying transformative data science and machine learning solutions to solve complex business problems in the telecom industry. He has notable achievements in developing a machine learning–driven customer churn prediction and root cause exploration solution, a customer credit scoring system, and a product recommendation engine.

Shitalkumar is skilled in enterprise data science and research ecosystem development, dedicated to optimizing key business indicators and adding revenue streams for companies. He is pursuing a doctorate in business administration from SSBM, Switzerland, and an MTech in computer science and engineering from VNIT Nagpur.

Shitalkumar has authored a book titled *Step Up for Leadership in Enterprise Data Science and Artificial Intelligence with Big Data: Illustrations with R and Python* and co-authored a book titled *Web Application Development with R Using Shiny, Third Edition*. He is a speaker at various technology and business events such as World AI Show Jakarta 2021, 2022, and 2023, NXT CX Jakarta 2022, Global Cloud-Native and Open Source Summit 2022, Cyber Security Summit 2022, and ASEAN Conversational Automation Webinar. You can find him on LinkedIn at www.linkedin.com/in/shitalkumars/.

Sandika S. Sukhdeve is an expert in data visualization and Google-certified project management. She previously served as Assistant Professor in the Mechanical Engineering Department and has authored Amazon bestseller titles across diverse markets such as the United States, Germany, Canada, and more. She has a background in human resources and a wealth of experience in branding.

As Assistant Professor, Sandika successfully guided more than 2,000 students, delivered 1,000+ lectures, and mentored numerous projects (including Computational Fluid Dynamics). She excels in managing both people and multiple projects, ensuring timely completion. Her areas of specialization encompass thermodynamics, applied thermodynamics, industrial engineering, product design and development, theory of machines, numerical methods and optimization, and fluid mechanics. She holds a master's degree in technology (with a specialization in heat power), and she possesses exceptional skills in visualizing, analyzing, and constructing classification and prediction models using R and MATLAB. You can find her on LinkedIn at `www.linkedin.com/in/sandika-awale/`.

About the Technical Reviewer

Sachin G. Narkhede is a highly skilled data scientist and software engineer with over 12 years of experience in Python and R programming for data analytics and machine learning. He has a strong background in building machine learning models using scikit-learn, Pandas, Seaborn, and NLTK, as well as developing question-answering machines and chatbots using Python and IBM Watson.

Sachin's expertise also extends to data visualization using Microsoft BI and the data analytics tool RapidMiner. With a master's degree in information technology, he has a proven track record of delivering successful projects, including transaction monitoring, trade-based money laundering detection, and chatbot development for banking solutions. He has worked on GCP (Google Cloud Platform).

Sachin's passion for research is evident in his published papers on brain tumor detection using symmetry and mathematical analysis. His dedication to learning is demonstrated through various certifications and workshop participation. Sachin's combination of technical prowess and innovative thinking makes him a valuable asset in the field of data science.

Acknowledgments

We extend our sincerest thanks to all those who have supported us throughout the writing process of this book. Your encouragement, guidance, and unwavering belief in our abilities have contributed to bringing this project to fruition.

Above all, we express our deepest gratitude to our parents, whose unconditional love, unwavering support, and sacrifices have allowed us to pursue our passions. Your unwavering belief in us has been the driving force behind our motivation.

We are grateful to our family for their understanding and patience during our countless hours researching, writing, and editing this book. Your love and encouragement have served as a constant source of inspiration.

A special thank you goes to our friends for their words of encouragement, motivation, and continuous support throughout this journey. Your belief in our abilities and willingness to lend an ear during moments of doubt have been invaluable.

We would also like to express our appreciation to our mentors and colleagues who generously shared their knowledge and expertise, providing valuable insights and feedback that have enriched the content of this book.

Lastly, we want to express our deepest gratitude to the readers of this book. Your interest and engagement in the subject matter make all our efforts worthwhile. We sincerely hope this book proves to be a valuable resource for your journey in understanding and harnessing the power of technology.

ACKNOWLEDGMENTS

Once again, thank you for your unwavering support, love, and encouragement. This book would not have been possible without each and every one of you.

Sincerely,

Shitalkumar and Sandika

Preface

The business landscape is transforming by integrating data science and machine learning, and cloud computing platforms have become indispensable for handling and examining vast datasets. Google Cloud Platform (GCP) stands out as a top-tier cloud computing platform, offering extensive services for data science and machine learning.

This book is a comprehensive guide to learning GCP for data science, using only the free-tier services offered by the platform. Regardless of your professional background as a data analyst, data scientist, software engineer, or student, this book offers a comprehensive and progressive approach to mastering GCP's data science services. It presents a step-by-step guide covering everything from fundamental concepts to advanced topics, enabling you to gain expertise in utilizing GCP for data science.

The book begins with an introduction to GCP and its data science services, including BigQuery, Cloud AI Platform, Cloud Dataflow, Cloud Storage, and more. You will learn how to set up a GCP account and project and use Google Colaboratory to create and run Jupyter notebooks, including machine learning models.

The book then covers big data and machine learning, including BigQuery ML, Google Cloud AI Platform, and TensorFlow. Within this learning journey, you will acquire the skills to leverage Vertex AI for training and deploying machine learning models and harness the power of Google Cloud Dataproc for the efficient processing of large-scale datasets.

The book then delves into data visualization and business intelligence, encompassing Looker Studio and Google Colaboratory. You will gain proficiency in generating and distributing data visualizations and reports using Looker Studio and acquiring the knowledge to construct interactive dashboards.

The book then covers data processing and transformation, including Google Cloud Dataflow and Google Cloud Dataprep. You will learn how to run data processing pipelines on Cloud Dataflow and how to use Cloud Dataprep for data preparation.

The book also covers data analytics and storage, including Google Cloud Storage, Google Cloud SQL, and Google Cloud Pub/Sub. You will learn how to use Cloud Pub/Sub for real-time data streaming and how to set up and consume data streams.

Finally, the book covers advanced topics, including securing and managing GCP resources with Identity and Access Management (IAM), using Google Cloud Source Repositories for version control, Dataplex, and Cloud Data Fusion.

Overall, this book provides a comprehensive guide to learning GCP for data science, using only the free-tier services offered by the platform. It covers the basics of the platform and advanced topics for individuals interested in taking their skills to the next level.

Introduction

Welcome to *Google Cloud Platform for Data Science: A Crash Course on Big Data, Machine Learning, and Data Analytics Services*. In this book, we embark on an exciting journey into the world of Google Cloud Platform (GCP) for data science. GCP is a cutting-edge cloud computing platform that has revolutionized how we handle and analyze data, making it an indispensable tool for businesses seeking to unlock valuable insights and drive innovation in the modern digital landscape.

As a widely recognized leader in cloud computing, GCP offers a comprehensive suite of services specifically tailored for data science and machine learning tasks. This book provides a progressive and comprehensive approach to mastering GCP's data science services, utilizing only the free-tier services offered by the platform. Whether you're a seasoned data analyst, a budding data scientist, a software engineer, or a student, this book equips you with the skills and knowledge needed to leverage GCP for data science purposes.

Chapter 1: "Introduction to GCP"

This chapter explores the transformative shift that data science and machine learning brought about in the business landscape. We highlight cloud computing platforms' crucial role in handling and analyzing vast datasets. We then introduce GCP as a leading cloud computing platform renowned for its comprehensive suite of services designed specifically for data science and machine learning tasks.

Chapter 2: "Google Colaboratory"

Google Colaboratory, or Colab, is a robust cloud-based platform for data science. In this chapter, we delve into the features and capabilities of Colab. You will learn how to create and run Jupyter notebooks, including machine learning models, leveraging Colab's seamless integration with GCP services. We also discuss the benefits of using Colab for collaborative data analysis and experimentation.

Chapter 3: "Big Data and Machine Learning"

This chapter explores the world of big data and machine learning on GCP. We delve into BigQuery, a scalable data warehouse, and its practical use cases. Next, we focus on BigQuery ML, which enables you to build machine learning models directly within BigQuery. We then focus on Google Cloud AI Platform, where you will learn to train and deploy machine learning models. Additionally, we introduce TensorFlow, a popular framework for deep learning on GCP. Lastly, we explore Google Cloud Dataproc, which facilitates the efficient processing of large-scale datasets.

Chapter 4: "Data Visualization and Business Intelligence"

Effective data visualization and business intelligence are crucial for communicating insights and driving informed decisions. In this chapter, we dive into Looker Studio, a powerful tool for creating and sharing data visualizations and reports. You will learn how to construct interactive dashboards that captivate your audience and facilitate data-driven decision-making. We also explore data visualization within Colab, enabling you to create compelling visual representations of your data.

Chapter 5: "Data Processing and Transformation"

This chapter emphasizes the importance of data processing and transformation in the data science workflow. We introduce Google Cloud Dataflow, a batch and stream data processing service. You will learn to design and execute data processing pipelines using Cloud Dataflow. Additionally, we cover Google Cloud Dataprep, a tool for data preparation and cleansing, ensuring the quality and integrity of your data.

Chapter 6: "Data Analytics and Storage"

Data analytics and storage play a critical role in data science. This chapter delves into Google Cloud Storage and its use cases for storing and managing data. Next, we explore Google Cloud SQL, a relational database service that enables efficient querying and analysis of structured data. Finally, we introduce Google Cloud Pub/Sub, a real-time data streaming service, and guide you through setting up and consuming data streams.

Chapter 7: "Advanced Topics"

In the final chapter, we tackle advanced topics to elevate your expertise in GCP for data science. We explore securing and managing GCP resources with Identity and Access Management (IAM), ensuring the confidentiality and integrity of your data. We also introduce Google Cloud Source Repositories, a version control system for your code. Lastly, we touch upon innovative technologies like Dataplex and Cloud Data Fusion, expanding your data science horizons.

By the end of this book, you will have gained comprehensive knowledge and practical skills in utilizing GCP for data science. You will be equipped with the tools and techniques to extract valuable insights from vast datasets, train and deploy machine learning models, and create impactful data visualizations and reports.

Moreover, this newfound expertise in GCP for data science will open up numerous opportunities for career advancement. Whether you're a data analyst, data scientist, software engineer, or student, the skills you acquire through this book will position you as a valuable asset in today's data-driven world. You can leverage GCP's robust services to develop end-to-end data solutions, seamlessly integrating them into the tech ecosystem of any organization.

So get ready to embark on this transformative journey and unlock the full potential of Google Cloud Platform for data science. Let's dive in together and shape the future of data-driven innovation!

CHAPTER 1

Introduction to GCP

Over the past few years, the landscape of data science has undergone a remarkable transformation in how data is managed by organizations. The rise of big data and machine learning has necessitated the storage, processing, and analysis of vast quantities of data for businesses. As a result, there has been a surge in the demand for cloud-based data science platforms like Google Cloud Platform (GCP).

According to a report by IDC, the worldwide public cloud services market was expected to grow by 18.3% in 2021, reaching $304.9 billion. GCP has gained significant traction in this market, becoming the third-largest cloud service provider with a market share of 9.5% (IDC, 2021). This growth can be attributed to GCP's ability to provide robust infrastructure, data analytics, and machine learning services.

GCP offers various data science services, including data storage, processing, analytics, and machine learning. It also provides tools for building and deploying applications, managing databases, and securing resources.

Let's look at a few business cases that shifted to GCP and achieved remarkable results:

1. **The Home Depot**: The Home Depot, a leading home improvement retailer, wanted to improve their online search experience for customers. They shifted their search engine to GCP and saw a 50% improvement in the speed of their search results. This led to a 10% increase in customer satisfaction and a 20% increase in online sales (Google Cloud, *The Home Depot*, n.d.).

S. R. Sukhdeve and S. S. Sukhdeve, *Google Cloud Platform for Data Science*, https://doi.org/10.1007/978-1-4842-9688-2_1

2. **Spotify**: Spotify, a popular music streaming service, migrated its data infrastructure to GCP to handle its growing user base. By doing so, Spotify reduced its infrastructure costs by 75% and handled more user requests per second (Google Cloud, *Spotify*, n.d.).

3. **Nielsen**: Nielsen, a leading market research firm, wanted to improve their data processing speed and accuracy. They shifted their data processing to GCP and achieved a 60% reduction in processing time, resulting in faster insights and better decision-making (Google Workspace, n.d.).

Apart from the preceding examples, other organizations have benefited from GCP's data science services, including Coca-Cola, PayPal, HSBC, and Verizon.

With GCP, data scientists can focus on solving complex business problems instead of worrying about managing infrastructure. For instance, let's consider the example of a retail company that wants to analyze customer behavior to improve its marketing strategy. The company has millions of customers and collects vast data from various sources such as social media, online purchases, and surveys. To analyze this data, the company needs a powerful cloud-based platform that can store and process the data efficiently. GCP can help the company achieve this by providing scalable data storage with Google Cloud Storage, data processing with Google Cloud Dataflow, and data analytics with Google BigQuery. The company can also leverage machine learning with Google Cloud AI Platform to build models that can predict customer behavior and Google Cloud AutoML to automate the machine learning process.

GCP's data science services offer valuable support to individuals pursuing careers in the field of data science. As the demand for data scientists continues to increase, possessing expertise in cloud-based platforms like GCP has become crucial. According to Glassdoor, data scientists in the United States earn an average salary of $113,309 per year. Additionally, a recent report by Burning Glass Technologies highlights a notable 67% increase in job postings that specifically require GCP skills within the past year. Consequently, acquiring knowledge in GCP can provide significant advantages for those seeking employment opportunities in the data science field.

This chapter provides an introduction to GCP along with its data science offerings, and it will walk you through the process of creating your GCP account and project.

Overview of GCP and Its Data Science Services

The architecture of Google Cloud Platform (GCP) is designed to provide users with a highly available, scalable, and reliable cloud computing environment. It is based on a distributed infrastructure that spans multiple geographic regions and zones, allowing for redundancy and failover in case of system failures or disasters (Sukhdeve, 2020).

At a high level, the architecture of GCP consists of the following layers:

1. **Infrastructure layer**: This layer includes the physical hardware and network infrastructure comprising the GCP data centers. It has servers, storage devices, networking equipment, and other components required to run and manage the cloud services.

3

2. **Compute layer**: This layer includes the virtual machines (VMs) and container instances that run user applications and services. It has services like Compute Engine, Kubernetes Engine, and App Engine.

3. **Storage layer**: This includes the various storage services GCP provides, such as Cloud Storage, Cloud SQL, and Cloud Bigtable. It offers scalable and durable storage options for multiple types of data.

4. **Networking layer**: This layer includes the various networking services provided by GCP, such as Virtual Private Cloud (VPC), Load Balancing, and Cloud CDN. It offers secure and reliable networking options for user applications and services.

5. **Management and security layer**: This includes the various management and security services GCP provides, such as Identity and Access Management (IAM), Security Command Center, and Stackdriver. It provides tools for managing and securing user applications and services running on GCP.

In this way, the architecture of GCP is designed to provide a flexible and customizable cloud computing environment that can meet the needs of a wide range of users, from small startups to large enterprises.

GCP is a suite of cloud computing services that runs on the same infrastructure that Google uses for its end user products, such as Google Search and YouTube (Google Cloud, 2023).

For data science, it offers several services, including

- **BigQuery**: A fully managed, serverless data warehousing solution

- **Cloud AI Platform**: A suite of machine learning tools, including TensorFlow, scikit-learn, and XGBoost

- **Cloud Dataflow**: A fully managed service for transforming and analyzing data

- **Cloud DataLab**: An interactive development environment for machine learning and data science

- **Cloud Dataproc**: A managed Apache Hadoop and Apache Spark service

- **Cloud Storage**: A scalable, fully managed object storage service

- **Cloud Vision API**: A pre-trained image analysis API

These services can be used together to build complete data science solutions, from data ingestion to model training to deployment. Many of the GCP data science services mentioned in the syllabus have a free tier, which provides limited access to the services for free. The following is a list of the services and their respective free-tier offerings:

- **Google Colaboratory**: Completely free.

- **BigQuery**: 1 TB of data processed per month for free.

- **Cloud AI Platform**: Free access to AI Platform Notebooks, which includes Colaboratory.

- **Looker Studio**: Completely free.

- **Cloud Dataflow**: Two free job hours per day.

- **Cloud Dataprep**: Free trial of two million records per month.

- **Cloud Storage**: 5 GB of standard storage per month for free.

- **Cloud SQL**: Free trial of second-generation instances, up to 125 MB of storage.

- **Cloud Pub/Sub**: The free tier includes one million operations per month.

Note   The free-tier offerings may be subject to change, and usage beyond the free tier will incur charges. It is recommended to check the GCP pricing page for the latest information on the free-tier offerings.

The various services of GCP can be categorized as follows:

1. **Compute**: This category includes services for running virtual machines (VMs) and containers, such as Compute Engine, Kubernetes Engine, and App Engine.

2. **Storage**: This category includes services for storing and managing different types of data, such as Cloud Storage, Cloud SQL, and Cloud Datastore.

3. **Networking**: This category includes services for managing networking resources, such as Virtual Private Cloud (VPC), Cloud Load Balancing, and Cloud DNS.

4. **Big data**: This category includes services for processing and analyzing large datasets, such as BigQuery, Cloud Dataflow, and Cloud Dataproc.

5. **Machine learning**: This category includes services for building and deploying machine learning models, such as Cloud AI Platform, AutoML, and AI Building Blocks.

6. **Security**: This category includes services for managing and securing GCP resources, such as Identity and Access Management (IAM), Cloud Key Management Service (KMS), and Cloud Security Command Center.

7. **Management tools**: This category includes services for managing and monitoring GCP resources, such as Stackdriver, Cloud Logging, and Cloud Monitoring.

8. **Developer tools**: This category includes services for building and deploying applications on GCP, such as Cloud Build, Cloud Source Repositories, and Firebase.

9. **Internet of Things (IoT)**: This category includes services for managing and analyzing IoT data, such as Cloud IoT Core and Cloud Pub/Sub.

Setting Up a GCP Account and Project

Here are the steps to set up a Google Cloud Platform (GCP) account and project:

- Go to the GCP website: https://cloud.google.com/.

- Click the Try it free button.

- Log in using your existing Google account, or create a new account if you do not have one.

- Fill out the required information for your GCP account.

- Once your account is set up, click the Console button to access the GCP Console.

- In the GCP Console, click the Projects drop-down menu in the top navigation bar.

- Click the New Project button.

- Enter a name and ID for your project, and select a billing account if you have multiple accounts.

- Click the Create button.

Your project is now set up, and you can start using GCP services.

Note   If you are using the free tier, make sure to monitor your usage to avoid charges, as some services have limitations. Also, you may need to enable specific services for your project to use them.

Summary

Google Cloud Platform (GCP) offers a comprehensive suite of cloud computing services that leverage the same robust infrastructure used by Google's products. This chapter introduced GCP, highlighting its essential services and their relevance to data science.

We explored several essential GCP services for data science, including BigQuery, Cloud AI Platform, Cloud Dataflow, Cloud DataLab, Cloud Dataproc, Cloud Storage, and Cloud Vision API. Each of these services serves a specific purpose in the data science workflow, ranging from data storage and processing to machine learning model development and deployment.

Furthermore, we discussed the availability of free-tier offerings for various GCP data science services, allowing users to get started and explore these capabilities at no cost. Staying updated with the latest information on free-tier offerings and pricing is important by referring to the GCP pricing page.

Following the outlined steps, users can set up their GCP account and create projects, providing access to a powerful cloud computing platform for their data science initiatives.

GCP's robust infrastructure, extensive range of services, and integration with popular data science tools make it a compelling choice for organizations and individuals looking to leverage the cloud for their data-driven projects. With GCP, users can harness Google's infrastructure's scalability, reliability, and performance to tackle complex data challenges and unlock valuable insights.

As we progress through this guide, we will delve deeper into specific GCP services and explore how they can be effectively utilized for various data science tasks.

CHAPTER 2

Google Colaboratory

Google Colaboratory is a free, cloud-based Jupyter Notebook environment provided by Google. It allows individuals to write and run code in Python and other programming languages and perform data analysis, data visualization, and machine learning tasks. The platform is designed to be accessible, easy to use, and collaboration-friendly, making it a popular tool for data scientists, software engineers, and students.

This chapter will guide you through the process of getting started with Colab, from accessing the platform to understanding its features and leveraging its capabilities effectively. We will cover how to create and run Jupyter notebooks, run machine learning models, and access GCP services and data from Colab.

Features of Colab

Cloud-based environment: Colaboratory runs on Google's servers, eliminating users needing to install software on their devices.

Easy to use: Colaboratory provides a user-friendly interface for working with Jupyter notebooks, making it accessible for individuals with limited programming experience.

Access to GCP services: Colaboratory integrates with Google Cloud Platform (GCP) services, allowing users to access and use GCP resources, such as BigQuery and Cloud Storage, from within the notebook environment.

© Shitalkumar R. Sukhdeve and Sandika S. Sukhdeve 2023
S. R. Sukhdeve and S. S. Sukhdeve, *Google Cloud Platform for Data Science*,
https://doi.org/10.1007/978-1-4842-9688-2_2

Sharing and collaboration: Colaboratory allows users to share notebooks and collaborate on projects with others, making it an excellent tool for team-based work.

GPU and TPU support: Colaboratory provides access to GPUs and TPUs for running computationally intensive tasks, such as deep learning and high-performance computing.

Google Colaboratory provides a powerful and flexible platform for data science, machine learning, and other technical tasks, making it a popular choice for individuals and teams looking to work with cloud-based tools.

Creating and Running Jupyter Notebooks on Colaboratory

To begin your journey with Colab, follow these steps to access the platform:

   i.   **Open a web browser**: Launch your preferred web browser on your computer or mobile device.

  ii.   **Sign up for a Google account**: Before using Colaboratory, you must have a Google account. If you don't have one, sign up for a free Google account (Figure 2-1).

Google

Sign in

Use your Google Account

Email or phone

|

Forgot email?

Not your computer? Use Guest mode to sign in privately.
Learn more

Create account   Next

Figure 2-1. *Sign up for a Google account*

iii. **Navigate to Colaboratory:** Visit the Google
Colaboratory website (`https://colab.research.`
`google.com/`) and log in with your Google account.
Upon logging in, you should see a dialog like the one
shown in Figure 2-2.

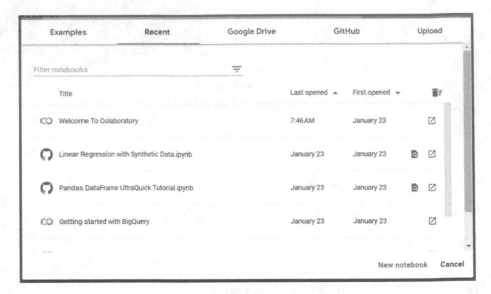

Figure 2-2. *Colaboratory after login*

iv. **Colab interface overview:**

Menu bar: The top section of the Colab interface houses the menu bar. It contains various options for file operations, runtime configuration, and more. Take a moment to explore the available menu options.

Toolbar: The toolbar is located directly below the menu bar. It provides quick access to common actions such as running, inserting new cells, and changing cell types.

Code cells: The main area of the Colab interface is composed of code cells. These cells allow you to write and execute Python code. By default, a new notebook starts with a single code cell. You can add additional cells as needed.

Text cells (Markdown cells): Besides code cells, Colab supports text cells, also known as Markdown cells. Text cells allow you to add explanatory text, headings, bullet points, and other formatted content to your notebook. They help provide documentation, explanations, and context to your code.

v. **Create a new notebook:** To create a new Colaboratory notebook, click the New notebook button in the top-right or bottom-right corner as shown in Figure 2-2.

vi. **Select the runtime type**: When creating a new notebook, you'll need to specify the runtime type. Select Python 3 to work with Python in Colaboratory.

vii. **Start coding:** You're now ready to start coding in Python. You can change the notebook's name by clicking the top-left corner where "Untitled1.ipynb" is written (Figure 2-3). Click the first code cell in your notebook to select it. Type a simple Python statement, such as print("Hello, Colab!"), into the code cell. Press Shift+Enter to execute the code. The output of the code will appear below the cell.

Figure 2-3. *New notebook*

15

Hands-On Example

Insert text in the notebook to describe the code by clicking the +
Text button.

Figure 2-4. *Insert text in the notebook*

Perform a sum of two numbers.

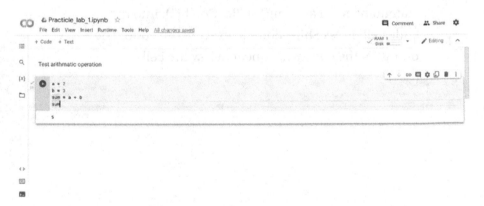

Figure 2-5. *Arithmetic example*

As can be seen from the preceding image, a = 2, b = 3, and sum = a + b,
giving an output of 5.

Here is an example in Python to generate random numbers using the random library and visualize the data using Matplotlib.

Figure 2-6. *Data science example*

The following code demonstrates how to generate random numbers, create a histogram, and visualize the data using the Python libraries random and matplotlib.pyplot:

```
import random
import matplotlib.pyplot as plt

# Generate 100 random numbers between 0 and 100
random_numbers = [random.randint(0,100) for i in range(100)]

# Plot the data as a histogram
plt.hist(random_numbers, bins=10)
plt.xlabel('Number')
plt.ylabel('Frequency')
plt.title('Random Number Distribution')
plt.show()
```

Here's a breakdown of the code:

Import random and import matplotlib.pyplot as plt: These lines import the necessary libraries for generating random numbers and creating visualizations using Matplotlib.

random_numbers = [random.randint(0,100) for i in range(100)]: This line generates a list called random_numbers containing 100 random integers between 0 and 100. The random.randint() function from the random library is used within a list comprehension to generate these random numbers.

plt.hist(random_numbers, bins=10): This line creates a histogram using the hist() function from matplotlib.pyplot. It takes the random_numbers list as input and specifies the number of bins (10) for the histogram. The hist() function calculates the frequency of values within each bin and plots the histogram.

plt.xlabel('Number'), plt.ylabel('Frequency'), plt.title('Random Number Distribution'): These lines set the labels for the x-axis and y-axis and the title of the plot, respectively.

plt.show(): This line displays the plot on the screen. The show() function from matplotlib.pyplot is called to render the histogram plot.

By executing this code in a Colaboratory notebook or a Python environment, you will see a histogram visualization like Figure 2-7 that shows the distribution of the randomly generated numbers. The x-axis represents the number range (0–100), and the y-axis represents the frequency of each number within the dataset.

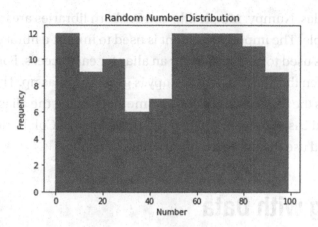

Figure 2-7. *Output random number distribution*

This example showcases the capability of Python and libraries like random and Matplotlib to generate and visualize data, providing a basic understanding of how to work with random numbers and histograms.

Importing Libraries

The following is an example code to import libraries into your notebook. If the library is not already installed, use the "!pip install" command followed by the library's name to install:

```
# Import the Pandas library
import pandas as pd
```

```
# Import the Numpy library
import numpy as np
```

```
# Import the Matplotlib library
import matplotlib.pyplot as plt
```

```
# Import the Seaborn library
import seaborn as sns
```

The Pandas, Numpy, Matplotlib, and Seaborn libraries are imported in this example. The import statement is used to import a library, and the as keyword is used to give the library an alias for easy access. For example, Pandas is given the alias pd, and Numpy is given the alias np. This allows you to access the library's functions and methods using the alias, which is shorter and easier to type. You can run this code in a Colaboratory notebook and use the imported libraries in your code.

Working with Data

Colaboratory integrates with Google Drive and Google Cloud Storage, allowing you to easily import and work with data from those services.

You can store or access data stored in Google Drive from Colaboratory by using the google.colab library. For example, to mount your Google Drive to Colaboratory, you can run the following code:

```
from google.colab import drive
    drive.mount('/content/drive')
```

Once you execute the code, a prompt will appear, asking you to sign into your Google account and grant Colaboratory the necessary permissions to access your Google Drive.

Once you've authorized Colaboratory, you can access your Google Drive data by navigating to /content/drive in the Colaboratory file explorer.

To write data to Google Drive, you can use Python's built-in open function. For example, to write a Pandas DataFrame to a CSV file in Google Drive, you can use the following code:

```
import pandas as pd
df = pd.DataFrame({'A': [1, 2, 3], 'B': [4, 5, 6]})
    df.to_csv('/content/drive/My Drive/Colab Notebooks/data.
    csv', index=False)
```

This code creates a new CSV file in the My Drive/Colab Notebooks/ folder in Google Drive with the data from the Pandas DataFrame.

Note that you can also use other libraries, such as gspread to write data to Google Sheets or google.cloud to write data to other GCP services. The approach you select will vary based on the desired data format and destination for writing.

You can use the data in your Google Drive just as you would any other data in Colaboratory. For example, you can read a CSV file stored in Google Drive and load it into a Pandas DataFrame using the following code:

```
import pandas as pd
df = pd.read_csv('/content/drive/My Drive/Colab
    Notebooks/data.csv')
```

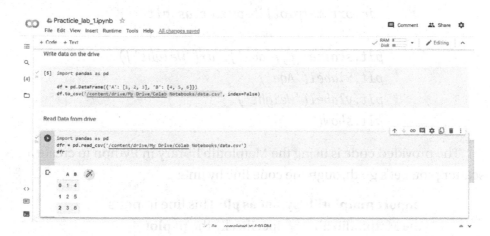

Figure 2-8. *Screenshot of Python on Colab*

Visualize Data

To visualize data in Colaboratory, you can use libraries such as Matplotlib or Seaborn to create plots and charts.

Create a data frame in Python using the Pandas library with the following code:

```
import pandas as pd
import numpy as np

df = pd.DataFrame({'age': np.random.randint(20, 80, 100),
                   'weight': np.random.randint(50, 100, 100)})
```

This will create a data frame with 100 rows and 2 columns named "age" and "weight", populated with random integer values between 20 and 80 for age and 50 and 100 for weight.

Visualize the data in the Panda's data frame using the Matplotlib library in Python. Here's an example to plot a scatter plot of the age and weight columns:

```
import matplotlib.pyplot as plt


plt.scatter(df['age'], df['weight'])
plt.xlabel('Age')
plt.ylabel('Weight')
plt.show()
```

The provided code is using the Matplotlib library in Python to create a scatter plot. Let's go through the code line by line:

> **import matplotlib.pyplot as plt**: This line imports the Matplotlib library, specifically the **pyplot** module. The **plt** is an alias or shorthand that allows us to refer to the module when calling its functions.

> **plt.scatter(df['age'], df['weight'])**: This line creates a scatter plot using the **scatter()** function from Matplotlib. It takes two arguments: **df['age']** and **df['weight']**. Assuming that **df** is a Pandas

DataFrame, this code is plotting the values from the "age" column on the x-axis and the values from the "weight" column on the y-axis.

plt.xlabel('Age'): This line sets the label for the x-axis of the scatter plot. In this case, the label is set to "Age."

plt.ylabel('Weight'): This line sets the label for the y-axis of the scatter plot. Here, the label is set to "Weight."

plt.show(): This line displays the scatter plot on the screen. It is necessary to include this line in order to see the plot rendered in a separate window or within the Jupyter notebook.

This will create a scatter plot with the age values on the x-axis and weight values on the y-axis. You can also use other types of plots, like histograms, line plots, bar plots, etc., to visualize the data in a Pandas data frame.

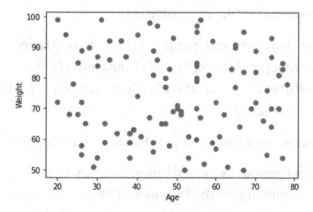

Figure 2-9. *Resultant graph age-weight distribution*

Running Machine Learning Models on Colaboratory

Here is an example of building a machine learning model using Python and the scikit-learn library. First, generate the data using the following code if you don't have one. The code generates random features and a random target array and then combines them into a Pandas DataFrame for further analysis or modeling:

```
import pandas as pd
import numpy as np

# Generate random features
np.random.seed(0)
n_samples = 1000
n_features = 5
X = np.random.randn(n_samples, n_features)

# Generate random target
np.random.seed(1)
y = np.random.randint(0, 2, n_samples)

# Combine the features and target into a data frame
df = pd.DataFrame(np.hstack((X, y[:, np.newaxis])),
columns=["feature_1", "feature_2", "feature_3", "feature_4",
"feature_5", "target"])
```

We will now examine the code step-by-step, analyzing each line:

1. **import pandas as pd** and **import numpy as np**: These lines import the Pandas and Numpy libraries, which are commonly used for data manipulation and analysis in Python.

2. **np.random.seed(0)**: This line sets the random seed for Numpy to ensure that the random numbers generated are reproducible. By setting a specific seed value (in this case, 0), the same set of random numbers will be generated each time the code is run.

3. **n_samples = 1000**: This line assigns the number of samples to be generated as 1000.

4. **n_features = 5**: This line assigns the number of features to be generated as 5.

5. **X = np.random.randn(n_samples, n_features)**: This line generates a random array of shape (n_samples, n_features) using the **np.random.randn()** function from Numpy. Each element in the array is drawn from a standard normal distribution (mean = 0, standard deviation = 1).

6. **np.random.seed(1)**: This line sets a different random seed value (1) for generating the random target values. This ensures that the target values are different from the features generated earlier.

7. **y = np.random.randint(0, 2, n_samples)**: This line generates an array of random integers between 0 and 1 (exclusive) of length n_samples using the **np.random.randint()** function. This represents the target values for classification, where each target value is either 0 or 1.

8. **df = pd.DataFrame(np.hstack((X, y[:,**
    np.newaxis])), columns=["feature_1", "feature_2",
    "feature_3", "feature_4", "feature_5", "target"]):
    This line combines the features (X) and target (y)
    arrays into a Pandas DataFrame. The **np.hstack()**
    function horizontally stacks the X and y arrays, and
    the resulting combined array is passed to the **pd.**
    DataFrame() function to create a DataFrame. The
    columns parameter is used to assign column names
    to the DataFrame, specifying the names of the
    features and the target.

Now, we can generate a machine learning model using the following
code. The code loads a dataset from a CSV file, splits it into training and
testing sets, trains a random forest classifier, makes predictions on the test
set, and evaluates the model's performance by calculating the accuracy:

```
from sklearn.model_selection import train_test_split
from sklearn.ensemble import RandomForestClassifier
from sklearn.metrics import accuracy_score

# Load the dataset
df = pd.read_csv("data.csv")

# Split the data into features (X) and target (y)
X = df.drop("target ", axis=1)
y = df["target "]

# Split the data into training and testing sets
X_train, X_test, y_train, y_test = train_test_split(X, y,
test_size=0.2)
```

```
# Train a Random Forest classifier
clf = RandomForestClassifier(n_estimators=100)
clf.fit(X_train, y_train)

# Make predictions on the test set
y_pred = clf.predict(X_test)

# Evaluate the model's performance
accuracy = accuracy_score(y_test, y_pred)
    print("Accuracy:", accuracy)
```

Let's go through the code line by line:

1. **from sklearn.model_selection import train_test_split**: This line imports the **train_test_split** function from the **model_selection** module in the scikit-learn library. This function is used to split the dataset into training and testing subsets.

2. **from sklearn.ensemble import RandomForestClassifier**: This line imports the **RandomForestClassifier** class from the **ensemble** module in scikit-learn. This class represents a random forest classifier, an ensemble model based on decision trees.

3. **from sklearn.metrics import accuracy_score**: This line imports the **accuracy_score** function from the **metrics** module in scikit-learn. This function is used to calculate the accuracy of a classification model.

4. **df = pd.read_csv("data.csv")**: This line reads a CSV file named "data.csv" using the **read_csv** function from the Pandas library. The contents of the CSV file are loaded into a Pandas DataFrame called **df**.

27

5. **X = df.drop("target", axis=1)**: This line creates a new DataFrame **X** by dropping the "target" column from the original DataFrame **df**. This DataFrame contains the features used for training the model.

6. **y = df["target"]**: This line creates a Series **y** by selecting the "target" column from the original DataFrame **df**. This Series represents the target variable or the labels corresponding to the features.

7. **X_train, X_test, y_train, y_test = train_test_split(X, y, test_size=0.2)**: This line splits the data into training and testing sets using the **train_test_split** function. The features **X** and the target **y** are passed as arguments, along with the **test_size** parameter, which specifies the proportion of the data to be used for testing (in this case, 20%).

8. **clf = RandomForestClassifier(n_estimators=100)**: This line creates an instance of the **RandomForestClassifier** class with 100 decision trees. The **n_estimators** parameter determines the number of trees in the random forest.

9. **clf.fit(X_train, y_train)**: This line trains the random forest classifier (**clf**) on the training data (**X_train** and **y_train**). The classifier learns patterns and relationships in the data to make predictions.

10. **y_pred = clf.predict(X_test)**: This line uses the trained classifier (**clf**) to make predictions on the test data (**X_test**). The predicted labels are stored in the **y_pred** variable.

11. **accuracy = accuracy_score(y_test, y_pred)**:
This line calculates the accuracy of the model by
comparing the predicted labels (**y_pred**) with the
true labels from the test set (**y_test**). The accuracy
score is stored in the **accuracy** variable.

12. **print("Accuracy:", accuracy)**: This line prints the
accuracy score on the console, indicating how well
the model performed on the test data.

Deploying the Model on Production

If you want to deploy a scikit-learn model, you can use the joblib library to
save the model and then serve the model using a REST API. The following
code installs the joblib and Flask libraries, saves a trained model to a file
using joblib, and sets up a Flask application to serve the model predictions.
The "/predict" endpoint is defined, and when a request is made to this
endpoint, the model is loaded, predictions are made, and the results are
returned as a JSON response.

You can use a cloud service such as Google Cloud Platform or Amazon
Web Services to host the REST API and make the predictions available to
your applications. Here's an example of how to do this:

1. **Install the joblib library:**

```
!pip install joblib
```

2. **Save the model:**

```
import joblib
model = clf # Your trained model
# Save the model
joblib.dump(model, '/content/drive/My
Drive/Colab Notebooks/model.joblib')
```

3. **Serve the model using Flask:**

```
!pip install flask

from flask import Flask, request
import joblib

app = Flask(__name__)

@app.route("/predict", methods=["POST"])
def predict():
    data = request.json[X_test]
model = joblib.load('/content/drive/My
Drive/Colab Notebooks/model.joblib')
predictions = model.predict(data)
print(predictions)
return {'predictions': predictions.tolist()}

if __name__ == "__main__":
app.run()
```

This example uses the Flask library to serve the model as a REST API. You can test the API by sending a POST request to the endpoint /predict with the input data in the request body.

We will now examine the code step-by-step, analyzing each line:

1. **!pip install joblib**: This line installs the joblib library using the pip package manager. joblib is a library in Python used for serialization and deserialization of Python objects, including models.

2. **import joblib**: This line imports the joblib module, which provides functions for saving and loading models.

3. **model = clf**: This line assigns the trained model
    (**clf**) to the **model** variable. It assumes that **clf** is the
    trained model object.

4. **joblib.dump(model, '/content/drive/My Drive/
    Colab Notebooks/model.joblib')**: This line saves
    the model to a file named "model.joblib" using the
    joblib.dump() function. The file will be saved in the
    specified directory path "/content/drive/My Drive/
    Colab Notebooks/."

5. **!pip install flask**: This line installs the Flask web
    framework using the pip package manager. Flask
    is a popular Python framework for building web
    applications.

6. **from flask import Flask, request**: This line imports
    the Flask class from the flask module. It also imports
    the request object, which allows accessing data from
    incoming HTTP requests.

7. **app = Flask(__name__)**: This line creates a Flask
    application instance named **app**.

8. **@app.route("/predict", methods=["POST"])**: This
    line decorates the following function to handle
    requests to the "/predict" endpoint with the HTTP
    POST method.

9. **def predict()::** This line defines a function named
    predict that will be executed when a request is
    made to the "/predict" endpoint.

10. **data = request.json[X_test]**: This line retrieves the
    JSON data from the request sent to the "/predict"
    endpoint. It assumes that the JSON data contains
    the "X_test" key.

11. **model = joblib.load('/content/drive/My Drive/ Colab Notebooks/model.joblib')**: This line loads the model from the file "model.joblib" using the joblib.load() function.

12. **predictions = model.predict(data)**: This line uses the loaded model to make predictions on the data obtained from the request.

13. **print(predictions)**: This line prints the predictions to the console.

14. **return {'predictions': predictions.tolist()}**: This line returns a JSON response containing the predictions as a list.

15. **if __name__ == "__main__":**: This line checks if the script is being run directly as the main module.

16. **app.run()**: This line starts the Flask development server to run the application.

Accessing GCP Services and Data from Colaboratory

Connecting to GCP: Colaboratory is a part of Google Cloud Platform (GCP), so you can easily access other GCP services and data from Colaboratory. To connect to GCP, you'll need to set up a Google Cloud project and grant Colaboratory access to your Google Cloud resources.

Authenticating with Google Cloud: To authenticate with Google Cloud, you'll need to run the following command in a Colaboratory cell:

```
!gcloud auth login
```

By executing this command, a web browser window will be launched, enabling you to log into your Google account and authorize Colaboratory to access your Google Cloud resources.

Accessing GCP services: You can use Colaboratory to access and work with other GCP services, such as Google Cloud Storage, Google BigQuery, and Google Cloud AI Platform.

Accessing GCP data: You can also use Colaboratory to access and work with data stored in GCP, such as data stored in Google Cloud Storage or Google BigQuery. You'll need to use Python libraries such as Pandas and the Google Cloud API to access this data.

Importing data from GCP: You can use Colaboratory to import data from other GCP services and store it in a Pandas DataFrame. As an example, you have the option to import data from Google BigQuery and store it within a DataFrame utilizing the capabilities of the BigQuery API.

Exporting data to GCP: You can also use Colaboratory to export data to other GCP services. For example, you can export a Pandas DataFrame to Google Cloud Storage or Google BigQuery.

The integration of Colab with other Google Cloud services is discussed in detail in upcoming chapters.

Summary

In this chapter, we learned about Google Colaboratory, a free cloud-based Jupyter Notebook environment offered by Google. We discovered its key features, including its cloud-based nature, user-friendly interface, integration with Google Cloud Platform (GCP) services, sharing and collaboration capabilities, and support for GPU and TPU computing. These features make Colaboratory an accessible and powerful tool for data scientists, software engineers, and students.

We explored the process of creating and running Jupyter notebooks on Google Colaboratory. We learned how to access the platform, sign up for a Google account, and navigate the Colab interface. We also learned how to create new Colaboratory notebooks, select the runtime type, and start coding in Python. Through hands-on examples, we practiced inserting text, performing arithmetic operations, generating random numbers, and visualizing data using libraries like Matplotlib and Seaborn.

Moreover, we delved into importing libraries into Colab notebooks and working with data stored in Google Drive using the google.colab library. We learned how to import data from and write data to Google Drive, as well as access data in Google Drive through Colaboratory. Additionally, we briefly touched upon running machine learning models on Google Colaboratory and mentioned the availability of powerful machine learning tools and libraries like TensorFlow and Keras. We even explored an example of building a machine learning model using the scikit-learn library.

Conclusively, this chapter provided an overview of Google Colaboratory and its features, guided us through the process of accessing and using Colab, and taught us various coding, data manipulation, and visualization techniques within the Colaboratory environment.

CHAPTER 3

Big Data and Machine Learning

Google Cloud Platform offers multiple services for the management of big data and the execution of Extract, Transform, Load (ETL) operations. Additionally, it provides tools for the training and deployment of machine learning models. Within this chapter, we will delve into BigQuery and its query execution. Moreover, we will gain insights into employing BigQuery ML for the development of machine learning models. The exploration extends to Google Cloud AI Platform, where we will gain practical experience with Vertex AI for training and deploying machine learning models.

As we progress, an introduction to Dataproc and its applications in data processing awaits us. The chapter culminates with an introduction to TensorFlow.

BigQuery

BigQuery is a serverless, highly scalable, cost-effective cloud data warehouse provided by Google Cloud Platform (GCP). It allows you to store and analyze large amounts of structured and semi-structured data using SQL-like queries. With BigQuery, you can run ad hoc queries on terabytes of data in seconds without the need for any setup or maintenance.

© Shitalkumar R. Sukhdeve and Sandika S. Sukhdeve 2023
S. R. Sukhdeve and S. S. Sukhdeve, *Google Cloud Platform for Data Science*,
https://doi.org/10.1007/978-1-4842-9688-2_3

BigQuery is suitable for various use cases, including

Data warehousing: BigQuery is often used as a data warehouse for storing and analyzing large amounts of data. To create a complete data analysis pipeline, you can integrate BigQuery with other GCP tools, such as Google Cloud Storage and Google Looker Studio.

Analytics: BigQuery is suitable for running large-scale analytics projects, such as business intelligence, predictive modeling, and data mining.

Machine learning: BigQuery offers the capability to train machine learning models and perform scalable predictions. It provides comprehensive support for TensorFlow and other popular machine learning frameworks.

Streaming data: BigQuery supports real-time data streaming, allowing you to process and analyze data in near real time.

Geospatial data analysis: BigQuery supports geospatial data analysis, allowing you to perform complex geographic calculations and visualization.

BigQuery's serverless architecture, scalability, and cost-effectiveness make it an attractive solution for organizations of all sizes, from small startups to large enterprises.

Running SQL Queries on BigQuery Data

Starting with BigQuery is straightforward and involves the following steps:

Set up a Google Cloud account: To use BigQuery, you must sign up for a Google Cloud account. You can start with a free trial and then choose a paid plan that meets your needs.

Create a project: Once you have a Google Cloud account, create a new project in the Google Cloud Console. This will be the home for your BigQuery data.

Load your data: You can load your data into BigQuery in several ways, including uploading CSV or JSON files, using a cloud-based data source such as Google Cloud Storage, or streaming data in real time.

Run a query: Once your data is loaded, you can start querying it using BigQuery's web UI or the command-line interface. BigQuery uses a SQL-like syntax, so if you're familiar with SQL, you'll find it easy to use.

Analyze your results: After running a query, you can view the results in the web UI or export them to a file for further analysis.

Visualize your data: You can also use tools like Google Looker Studio to create visualizations and reports based on your BigQuery data.

BigQuery provides detailed documentation and tutorials to help you get started. Additionally, GCP provides a free community support forum, and if you have a paid plan, you can also access technical support.

Before jumping to the production environment and using a credit card, you can try BigQuery Sandbox.

BigQuery Sandbox is a free, interactive learning environment provided by Google Cloud Platform (GCP) to help users explore and learn about BigQuery. With BigQuery Sandbox, you can run queries on public datasets and test out the functionality of BigQuery without incurring any charges.

BigQuery Sandbox presents a user-friendly, web-based interface that facilitates SQL-like queries on BigQuery data. It offers a hassle-free experience, eliminating the need for any setup or installation. This makes it an invaluable resource for individuals such as data analysts, data scientists, and developers who are new to BigQuery and seek a hands-on learning experience with the tool while exploring its capabilities.

BigQuery Sandbox is not meant for production use and has limitations, such as a smaller data size limit and slower performance than a fully configured BigQuery instance. However, it's a great way to get started with BigQuery and gain familiarity with its functionality before using it in a production environment.

Here's a step-by-step tutorial for using BigQuery Sandbox:

Open the BigQuery Sandbox website: Go to the BigQuery Sandbox website at `https://console.cloud.google.com/bigquery`.

Figure 3-1. *The console of Google Cloud for BigQuery*

Log into your Google account: If you do not currently possess a Google account, it will be necessary to create one. Subsequently, you can log into your account in order to gain access to the BigQuery Sandbox environment. Follow the prompts to create a Google Cloud project. To use BigQuery Sandbox, you must create a Google Cloud project. It can be seen in the following image.

New Project

> ⚠ You have 12 projects remaining in your quota. Request an increase or delete projects. Learn more
>
> MANAGE QUOTAS

Project name *
Practice Project 1   ❓

Project ID: practice-project-1-377008. It cannot be changed later. EDIT

Location *
🏢 No organisation   BROWSE
Parent organisation or folder

CREATE   CANCEL

Figure 3-2. *Create a new project*

After creating the project, you can find the following screen.

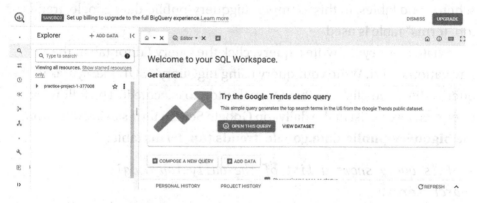

Figure 3-3. *Screen after creating a new project*

Explore the available datasets: BigQuery Sandbox provides access to several public datasets that you can use to practice running queries. To see a list of available datasets, click the Datasets tab in the navigation menu on the left-hand side of the screen.

From the preceding screen, you can see that you can try the Google Trends demo query. After clicking the OPEN THIS QUERY button, you can see the following results.

Figure 3-4. Google Trend demo query

Select a dataset: Click the name of a dataset to select it and see its schema and tables. In this example "bigquery-public-data.google_trends.top_terms" table is used.

Write a query: To write a query, click the Query Editor tab in the navigation menu. Write your query using BigQuery's SQL-like syntax in the query editor and click the Run Query button to execute it. The following query retrieves a list of the daily top Google Search terms using data from the **bigquery-public-data.google_trends.top_terms** table:

```
-- This query shows a list of the daily top Google
Search terms.
SELECT
    refresh_date AS Day,
    term AS Top_Term,
        -- These search terms are in the top 25 in the US
        each day.
    rank,
FROM `bigquery-public-data.google_trends.top_terms`
WHERE
    rank = 1
        -- Choose only the top term each day.
```

```
AND refresh_date >= DATE_SUB(CURRENT_DATE(),
INTERVAL 2 WEEK)
      -- Filter to the last 2 weeks.
GROUP BY Day, Top_Term, rank
ORDER BY Day DESC
      -- Show the days in reverse chronological order.
```

Here's a breakdown of the query:

The **SELECT** statement specifies the columns to be included in the result set. It selects the **refresh_date** column and aliases it as "Day," the **term** column as "Top_Term," and the **rank** column.

The **FROM** clause specifies the table **bigquery-public-data.google_trends.top_terms** from which the data will be retrieved.

The **WHERE** clause filters the data based on specific conditions. In this case, it includes only the rows where the **rank** column is equal to 1, indicating the top term for each day. It also filters the data to include only the records from the past two weeks by comparing the **refresh_date** column with the result of **DATE_SUB(CURRENT_DATE(), INTERVAL 2 WEEK)**, which subtracts two weeks from the current date.

The **GROUP BY** clause groups the result set by the "Day," "Top_Term," and "rank" columns. This ensures that each combination of these columns appears only once in the result set.

The **ORDER BY** clause orders the result set in descending order based on the "Day" column. This means that the most recent days will appear first in the result.

Concisely, the query retrieves the top Google Search term for each day over the past two weeks and presents them in reverse chronological order, first showing the most recent days.

View the query results: You'll see the results in the Results tab once your query has been executed. You can also save your query by clicking the Save button.

Repeat the process: Feel free to iterate through this process multiple times, experimenting with various datasets and crafting diverse queries to delve into the extensive capabilities offered by BigQuery.

Explore data: With Query results, there are options to save results and explore data.

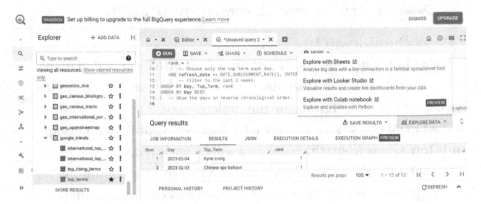

Figure 3-5. *Explore data with BigQuery*

Explore data with Colab: Once you click the Explore with Colab Notebook, you will be able to see the following screen.

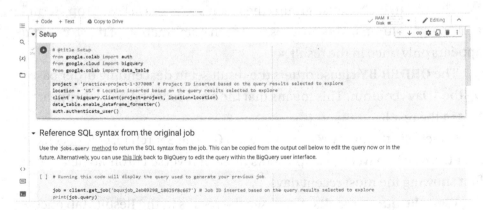

Figure 3-6. *Colab sheet*

Follow the same process as followed in using Colab in Chapter 2 and start exploring the data.

Explore data with Looker Studio: Once you click Explore with Looker Studio, the following screen can be seen.

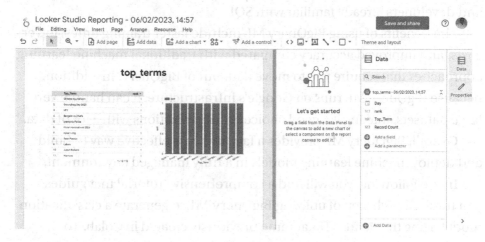

Figure 3-7. *Data exploration with Looker Studio*

BigQuery Sandbox is a simple and intuitive tool that provides a great way to get started with BigQuery and gain familiarity with its functionality. The web-based interface and public datasets make it easy to start running queries and learning about the tool.

BigQuery ML

BigQuery ML is a feature in Google Cloud's BigQuery that allows you to build machine learning models directly within BigQuery, without the need to move data out of BigQuery. This means you can use BigQuery ML to train machine learning models on large datasets that you have stored in BigQuery.

BigQuery ML supports several types of machine learning models, including linear regression, logistic regression, k-means clustering, and matrix factorization. You can use the SQL syntax in BigQuery to create and evaluate your machine learning models, making it easy for data analysts and developers already familiar with SQL.

The benefits of using BigQuery ML include faster training times, lower costs, and improved accuracy compared with traditional machine learning approaches that require you to move data out of BigQuery. In addition, because BigQuery ML runs on Google's infrastructure, it can handle very large datasets, making it a good choice for organizations with a lot of data.

Overall, BigQuery ML provides a fast and cost-effective way to build and deploy machine learning models in a fully managed environment.

In the following, you will find a comprehensive tutorial that guides you through each step of utilizing BigQuery ML to generate a classification model using the Pandas DataFrame previously created in Colab. To regenerate the dataset, execute the provided code on Colab:

```
import pandas as pd
import numpy as np

# Generate random features
np.random.seed(0)
n_samples = 1000
n_features = 5
X = np.random.randn(n_samples, n_features)

# Generate random target
np.random.seed(1)
y = np.random.randint(0, 2, n_samples)

# Combine the features and target into a data frame
```

```
df = pd.DataFrame(np.hstack((X, y[:, np.newaxis]))),
columns=["feature_1", "feature_2", "feature_3", "feature_4",
"feature_5", "target"])

# Save the data to a CSV file
df.to_csv("data.csv", index=False)
```

You can find the data.csv in the Colab files. Download the file and save it on your computer.

Create a BigQuery dataset: Go to the Google Cloud Console, select BigQuery, and create a new BigQuery dataset.

Upload the data to BigQuery: Go to the BigQuery Editor and create a new table. Then, upload the data to this table by clicking Create Table and selecting Upload. Choose the file containing the data and select the appropriate options for the columns.

Figure 3-8. *Create Table on BigQuery*

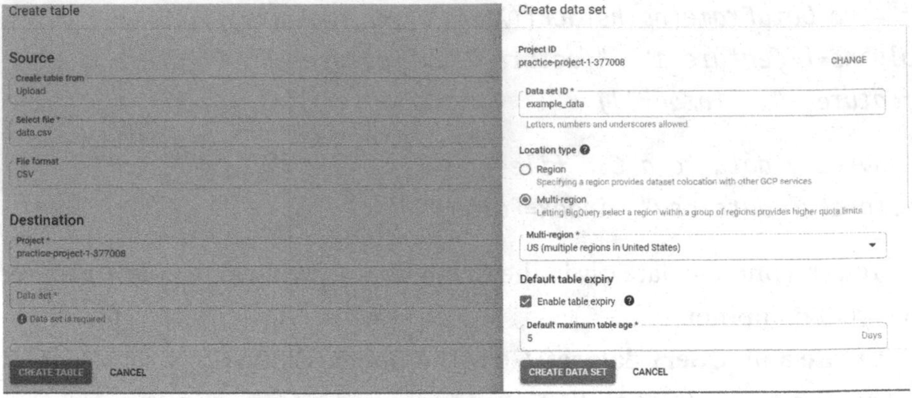

***Figure 3-9.** Create Table on BigQuery—setting various options*

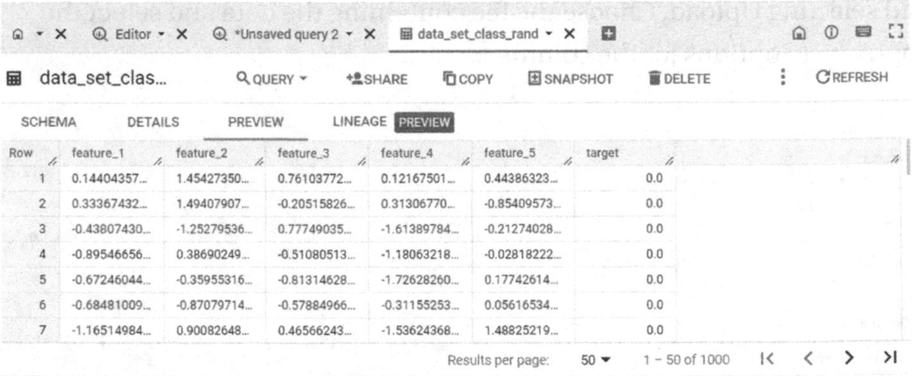

Row	feature_1	feature_2	feature_3	feature_4	feature_5	target
1	0.14404357...	1.45427350...	0.76103772...	0.12167501...	0.44386323...	0.0
2	0.33367432...	1.49407907...	-0.20515826...	0.31306770...	-0.85409573...	0.0
3	-0.43807430...	-1.25279536...	0.77749035...	-1.61389784...	-0.21274028...	0.0
4	-0.89546656...	0.38690249...	-0.51080513...	-1.18063218...	-0.02818222...	0.0
5	-0.67246044...	-0.35955316...	-0.81314628...	-1.72628260...	0.17742614...	0.0
6	-0.68481009...	-0.87079714...	-0.57884966...	-0.31155253...	0.05616534...	0.0
7	-1.16514984...	0.90082648...	0.46566243...	-1.53624368...	1.48825219...	0.0

***Figure 3-10.** After the table upload, preview the table*

Create a BigQuery ML model: In the BigQuery Editor, create a new SQL query. Use the CREATE MODEL statement to create a BigQuery ML model. Here's an example:

```
CREATE MODEL your_model_name
OPTIONS (model_type='logistic_reg', input_label_
cols=['target']) AS
SELECT *
FROM your_dataset_name.your_table_name
```

In the preceding syntax, enter *your_model_name* and *your_dataset_name.your_table_name* in (` `). See the following code for the uploaded dataset:

```
CREATE MODEL `example_data.data_set_class_rand_model`
OPTIONS (model_type='logistic_reg', input_label_
cols=['target']) AS
SELECT *
FROM `practice-project-1-377008.example_data.data_set_
class_rand`
```

Here, the model name is `example_data.data_set_class_rand_model` and the table name is `practice-project-1-377008.example_data.data_set_class_rand`. For finding the table name, click the table and go to the DETAILS option. The following screen will appear. Find Table ID and copy it and paste it into the placeholder *your_dataset_name.your_table_name*.

Figure 3-11. *Find the table name*

Now, your editor should look like this. If everything is fine, you will be able to find a green tick in the right corner like the following. Go ahead and run the query.

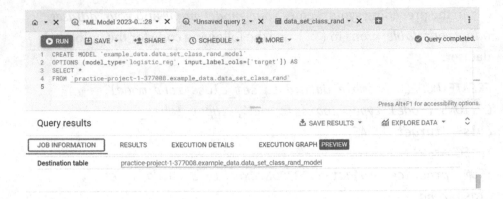

Figure 3-12. *BigQuery ML*

Go to Results and click Go to model. For model evaluation, the following results can be seen.

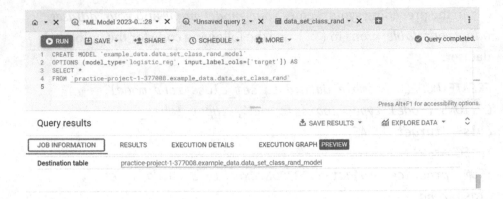

Figure 3-13. *Model evaluation after training*

Evaluate the model: In the BigQuery Editor, create a new SQL query. Use the ML.EVALUATE function to evaluate the model. Here's an example of how to do this:

```
SELECT *
FROM ML.EVALUATE(MODEL your_model_name, (
  SELECT *
  FROM your_dataset_name.your_table_name
))
```

In the preceding syntax, put all the details as before, and your query should look like the following:

```
SELECT *
FROM ML.EVALUATE(MODEL `example_data.data_set_class_rand_
model`, (
  SELECT *
  FROM `practice-project-1-377008.example_data.data_set_
  class_rand`
))
```

Run the query and you can find the following results.

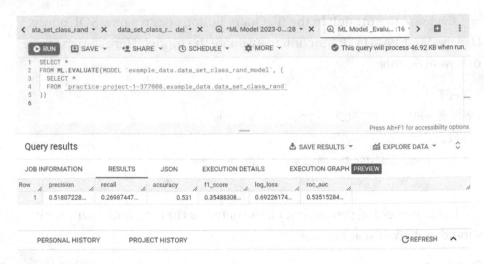

Figure 3-14. *Model evaluation*

Make predictions: In the BigQuery Editor, create a new SQL query. Use the ML.PREDICT function to make predictions. Here's an example of how to do this:

```
SELECT *
FROM ML.PREDICT(MODEL your_model_name, (
  SELECT *
  FROM your_dataset_name.your_table_name
))
```

Again insert all the information as before, and your query should look like the following:

```
SELECT *
FROM ML.PREDICT(MODEL `example_data.data_set_class_rand_
model`, (
  SELECT *
  FROM `practice-project-1-377008.example_data.data_set_
  class_rand`
))
```

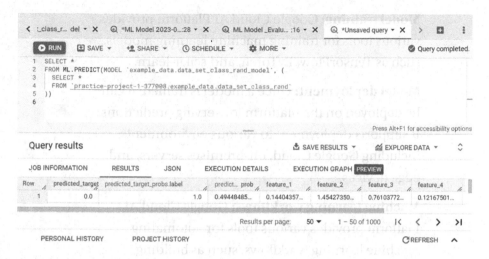

Figure 3-15. Prediction

This is a high-level overview of using BigQuery ML to create a classification model using the BigQuery Editor. For a more detailed explanation of the different functions and options, please refer to the BigQuery ML documentation.

Google Cloud AI Platform and Its Capabilities

Google Cloud AI Platform is an advanced cloud-native platform designed for the development, deployment, and management of machine learning models. It offers a comprehensive range of tools and services that facilitate the training and deployment of machine learning models, along with features for monitoring and optimizing their performance. Among its notable capabilities are

Model training: Google Cloud AI Platform provides various tools for training machine learning models, such as TensorFlow, PyTorch, and scikit-learn.

Model deployment: Once a model is trained, it can be deployed on the platform for serving predictions. It supports deployment to various environments, including Google Cloud, on-premises servers, and edge devices.

Machine learning workflows: Google Cloud AI Platform provides various tools for automating machine learning workflows, such as building, training, and deploying models.

Monitoring and management: Google Cloud AI Platform provides tools for monitoring the performance of deployed models and managing the resources used by the models.

Collaboration and sharing: The platform provides features for sharing and collaborating on machine learning models, including version control, team management, and access control.

Conclusively, Google Cloud AI Platform provides a comprehensive solution for building, deploying, and managing machine learning models at scale.

Here is a step-by-step tutorial on how to use Google Cloud AI Platform:

Set up a Google Cloud account: If you don't already have a Google Cloud account, create one. This will give you access to Google Cloud AI Platform and all of its other services.

Create a project: In the Google Cloud Console, create a new project for your machine learning model. This will be the place where you store your model and all of its related resources.

Upload your data: Upload your data to Google Cloud Storage. This is where your machine learning model will store and access the data.

Create a training job: Go to the AI Platform section of the Google Cloud Console, and create a new training job. Select the type of model you want to train and configure the training parameters, such as the number of epochs and the batch size.

Start the training job: Start the training job by clicking the Train button. The training job will run in the cloud and use your data to train the machine learning model.

Monitor the training job: You can monitor the progress of the training job from the AI Platform section of the Google Cloud Console. This will show you how the model's accuracy is improving over time.

Evaluate the model: Upon the completion of the training job, it is possible to assess its performance by evaluating various metrics, including accuracy, precision, and recall. This allows for a comprehensive understanding of how well the model performs.

Deploy the model: Once you're satisfied with the model's performance, you can deploy it to Google Cloud AI Platform. This will allow you to serve predictions from the model to your users.

Monitor the deployed model: Google Cloud AI Platform provides tools for monitoring the performance of deployed models. These tools can be utilized to verify the model's performance against the desired expectations and make any required modifications accordingly.

Maintain the model: Over time, your model may need to be retrained or updated as new data becomes available. Google Cloud AI Platform provides tools for updating and maintaining your model so that it continues to perform well.

The provided information offers a broad overview of the usage of Google Cloud AI Platform. However, please note that the specific steps and intricacies of the process may vary depending on the type of model being constructed and the data being utilized.

Google Cloud AI Platform offers several tools for interacting with its services:

Google Cloud Console: The console serves as a web-based graphical user interface for effectively managing your Google Cloud resources, which include AI Platform. It enables you to effortlessly create, manage, and monitor various AI Platform services and resources, such as AI models and datasets.

Google Cloud SDK: A command-line interface for managing your Google Cloud resources, including AI Platform. You can use the SDK to automate common tasks, such as deploying AI models and managing resources.

REST API: A programmatic interface for interacting with AI Platform using HTTP requests. The REST API allows you to programmatically access AI Platform services and resources, making it easy to integrate AI Platform into your existing workflows.

Google Cloud client libraries: Pre-written code libraries for accessing AI Platform services in popular programming languages, such as Python and Java. The client libraries simplify the process of integrating with AI Platform and make it easier to build applications that use AI Platform services.

Jupyter Notebook: A web-based interactive environment for writing and running code. You can use Jupyter Notebook to explore and analyze data, build and test AI models, and perform other data science tasks. Jupyter Notebook integrates with AI Platform, making it easy to work with AI Platform services and resources.

Google Cloud AI Platform Notebooks: A managed Jupyter Notebook environment that runs on Google Cloud. With AI Platform Notebooks, you can easily access and run AI Platform services and resources without having to set up your Jupyter Notebook environment.

Vertex AI Workbench: A web-based integrated development environment (IDE) for building and deploying machine learning models on Google Cloud. Vertex AI Workbench provides a user-friendly interface for building and training models and supports popular machine learning frameworks like TensorFlow and PyTorch.

Deep Learning VM: A pre-configured virtual machine image that provides a complete environment for developing and deploying deep learning models. The Deep Learning VM includes popular deep learning frameworks like TensorFlow and PyTorch, as well as a range of tools for data processing and analysis.

With a diverse set of tools at your disposal, you can interact with AI Platform in a manner that suits your individual and team needs. These tools offer various options, enabling you to select the most suitable one based on your specific requirements and personal preferences.

Using Vertex AI for Training and Deploying Machine Learning Models

Vertex AI, developed by OpenAI, is an inclusive platform to construct and launch machine learning models. It streamlines the process for data scientists, engineers, and developers by abstracting away the complexities of underlying technologies. Vertex AI offers a user-friendly interface that simplifies working with machine learning models and incorporates pre-built models and tools for essential tasks like data preparation, model selection, training, and deployment. Leveraging popular machine learning frameworks like PyTorch and TensorFlow, Vertex AI facilitates model development across diverse applications, including image classification, natural language processing, and recommendation systems.

It replaces several legacy services, including

> **Cloud AutoML:** As mentioned earlier, Cloud AutoML is being replaced by Vertex AutoML. Vertex AutoML provides a more streamlined and integrated experience for building and deploying machine learning models.

> **Cloud ML Engine:** The Vertex AI platform is replacing Cloud ML Engine. The Vertex AI platform offers functionality similar to Cloud ML Engine but with a more user-friendly interface and advanced features.

> **Cloud AI Platform Notebooks:** Cloud AI Platform Notebooks is being replaced by Vertex AI Notebooks. Vertex AI Notebooks provides the same functionality as Cloud AI Platform Notebooks but with more powerful hardware options and improved integration with other Vertex AI services.

Google Cloud previously provided AutoML Tables, a machine learning service enabling users to generate custom models without extensive programming or data science expertise. However, the legacy edition of AutoML Tables has been phased out and will no longer be accessible on Google Cloud starting from January 23, 2024.

The good news is that all the features and functionalities previously available in the legacy AutoML Tables, along with new additions, can now be accessed on the Vertex AI platform.

The Vertex AI platform integrates with the machine learning (ML) workflow in several key ways:

Data preparation: Vertex AI provides tools for preparing and cleaning your data for use in a machine learning model. This includes data loading, splitting, and preprocessing.

- Leverage the capabilities of Vertex AI Workbench notebooks to explore and visualize data effectively. Vertex AI Workbench seamlessly integrates with Cloud Storage and BigQuery, providing expedited access and processing of your data.

- For large datasets, use Dataproc Serverless Spark from a Vertex AI Workbench notebook to run Spark workloads without having to manage your Dataproc clusters (Google Cloud, 2023).

Model selection: Within Vertex AI, you will find a collection of pre-built models catering to common machine learning tasks like image classification, natural language processing, and recommendation systems. Additionally, AutoML can be employed to train models without the need for manual code writing (Blog, Neos Vietnam, n.d.). These models can be used as is or customized as needed. Additionally, Vertex AI provides the ability to create custom models from scratch.

Training: Vertex AI provides tools for training machine learning models on your data. The training process is managed within the platform, and Vertex AI provides monitoring and early-stopping capabilities to ensure that models are trained efficiently and effectively.

Model evaluation: Vertex AI provides tools for evaluating the performance of trained models. This includes performance metrics, such as accuracy, precision, and recall, as well as visualization tools to help you understand how your model is performing.

Deployment: Vertex AI provides tools for deploying trained models into production, making it possible to use your models to make predictions on new data. Vertex AI provides a number of deployment options, including serving the model as a REST API, deploying it as a function on a cloud platform, or integrating it into an existing application.

By integrating with the machine learning workflow in these key ways, Vertex AI makes it possible for data scientists, engineers, and developers to build, train, and deploy machine learning models more easily and efficiently.

MLOps, or Machine Learning Operations, refers to the process of managing and scaling machine learning models in a production environment. Vertex AI provides support for MLOps by offering several tools and features to help manage the deployment and operation of machine learning models. Some of the key features for MLOps on Vertex AI include

1. **Deployment**: Vertex AI provides tools for deploying machine learning models into production, making it possible to use your models to make predictions on new data. Vertex AI provides several deployment options, including serving the model as a REST API, deploying it as a function on a cloud platform, or integrating it into an existing application.

2. **Monitoring**: Vertex AI provides tools for monitoring the performance of deployed models. This includes performance metrics, such as accuracy, precision, and recall, as well as visualization tools to help you understand how your model is performing in real-world use.

3. **Version control:** Vertex AI provides version control capabilities, allowing you to manage multiple versions of a model and track changes over time. This makes it possible to roll back to previous versions if needed and helps ensure that you can maintain control over your models as they evolve.

4. **Scale**: Vertex AI provides tools for scaling machine learning models, making it possible to handle increasing amounts of data and traffic. This includes the ability to deploy models across multiple GPUs or cloud instances and to adjust the resources allocated to a model as needed.

By offering these and other features for MLOps, Vertex AI helps simplify the process of deploying and managing machine learning models in a production environment. This makes it easier for data scientists, engineers, and developers to focus on the important work of developing and refining their models while still maintaining control and visibility over the models as they are used in the real world.

To use Vertex AI, you must opt for either a free tier or a full-service account. Once done, you can click the link `https://console.cloud.google.com/` and access the console.

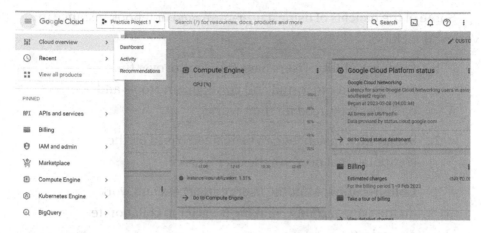

Figure 3-16. *Console*

Now you are ready to access Google Cloud services. Go to the search box and type Vertex AI, and you will be able to see the following screen.

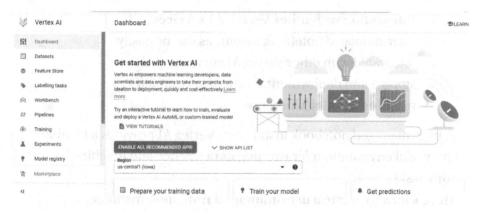

Figure 3-17. *Vertex AI Dashboard*

A user-managed notebook instance in Vertex AI provides a dedicated and customizable environment for running data science and machine learning tasks. The main use of this type of instance is to provide data scientists and machine learning engineers with a self-contained and scalable environment to develop and experiment with their models.

61

Some of the benefits of using a user-managed notebook instance in Vertex AI include

1. **Customizability**: Users have full control over the instance and can install and configure the tools and packages they need for their work.

2. **Scalability**: Instances can be resized based on the computational needs of the user's tasks.

3. **Collaboration**: Multiple users can access and work on the same instance, making it easier to collaborate on projects.

4. **Portability**: Instances can be saved and exported, allowing users to move their work to different platforms or share it with others.

5. **Integration with other Vertex AI services**: User-managed notebook instances can be easily integrated with other Vertex AI services, such as storage and deployment, to create a complete machine learning workflow.

A user-managed notebook instance in Vertex AI provides a flexible and powerful environment for running data science and machine learning tasks.

Here's how to create a user-managed notebook instance.

Here is the step-by-step method of creating a user-managed notebook instance.

1. Create a new project or select an existing project in which you have the relevant permissions.

Figure 3-18. *Selecting a project*

2. Enable the Notebooks API.

Figure 3-19. *Enabling the Notebook API*

3.   Start Vertex AI Workbench.

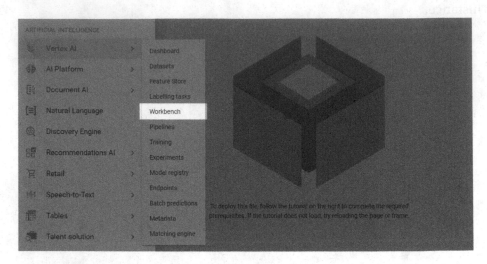

Figure 3-20. *Starting Vertex AI Workbench*

4.   Navigate to Vertex AI Workbench user-managed notebooks.

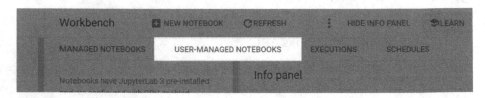

Figure 3-21. *User-managed notebooks*

5. Create an instance.

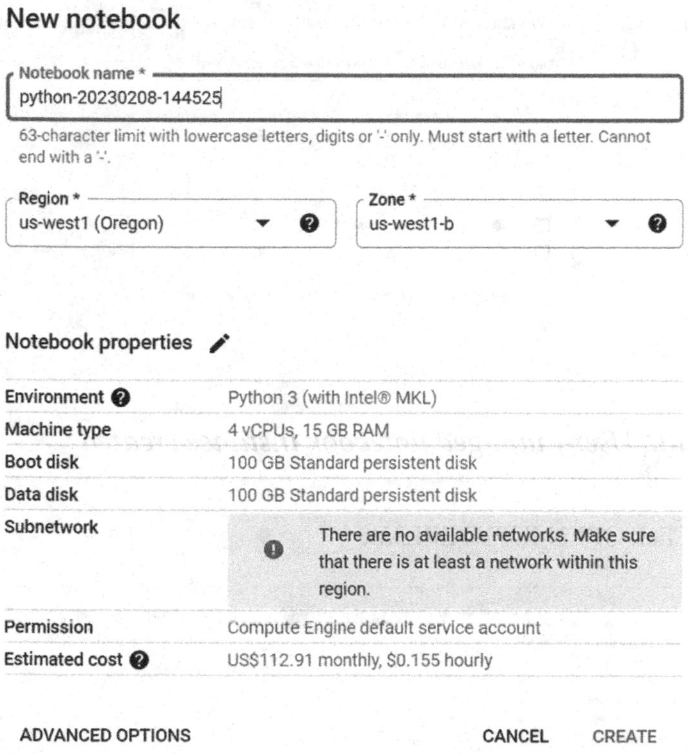

New notebook

Notebook name *
python-20230208-144525

63-character limit with lowercase letters, digits or '-' only. Must start with a letter. Cannot end with a '-'.

Region *
us-west1 (Oregon)   ▼   ❷

Zone *
us-west1-b   ▼   ❷

Notebook properties ✎

Environment ❷	Python 3 (with Intel® MKL)
Machine type	4 vCPUs, 15 GB RAM
Boot disk	100 GB Standard persistent disk
Data disk	100 GB Standard persistent disk
Subnetwork	There are no available networks. Make sure that there is at least a network within this region.
Permission	Compute Engine default service account
Estimated cost ❷	US$112.91 monthly, $0.155 hourly

ADVANCED OPTIONS   CANCEL   CREATE

Figure 3-22. *Create a new notebook instance*

You should be able to see the following window.

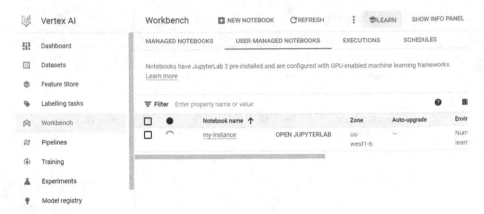

Figure 3-23. *User-managed notebook instance created*

6.   Click OPEN JUPYTERLAB.

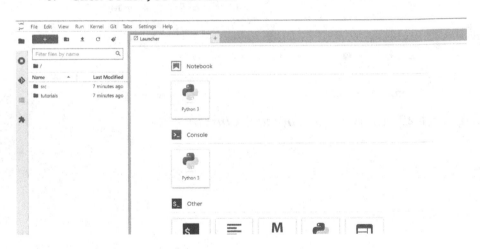

Figure 3-24. *Open a new Jupyter instance*

A **Cloud Storage bucket** is a container for storing data in the cloud. It is a central repository that can be used to store and access data from multiple locations and applications.

In Vertex AI, Cloud Storage buckets are used to store data required for training machine learning models and store the models themselves once they are deployed. Some common use cases for Cloud Storage buckets in Vertex AI include

1. **Storing training data**: Datasets that are used for training machine learning models can be stored in a Cloud Storage bucket for easy access and management.

2. **Saving model outputs**: The outputs from trained models, such as trained weights and model artifacts, can be kept in a Cloud Storage bucket for future use.

3. **Serving models**: Deployed models can be stored in a Cloud Storage bucket and served to end users through a REST API.

4. **Collaboration**: Collaboration on projects becomes more convenient with the ability to share Cloud Storage buckets among multiple users.

Cloud Storage buckets play a vital role in the machine learning workflow within Vertex AI, serving as a centralized repository for storing and retrieving data and models. They are an indispensable component in the smooth functioning of the machine learning process.

Here's how to create a Cloud Storage bucket:

1. Open the Google Cloud Console navigation menu and then select Cloud Storage.

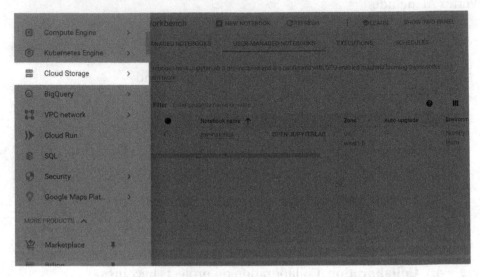

Figure 3-25. *Cloud Storage*

2. Click CREATE.

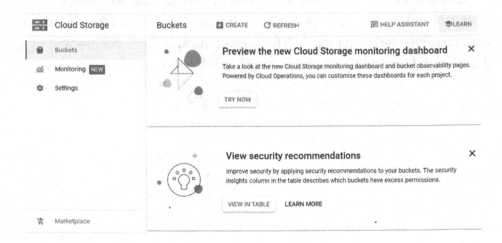

Figure 3-26. *Buckets*

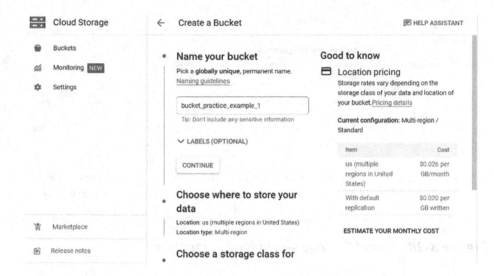

Figure 3-27. *New bucket creation*

3. Enter a unique name for the bucket, select the
    desired region, and choose the desired storage class,
    such as standard or infrequent access. Configure
    any additional settings, such as versioning and
    lifecycle rules. Click the Create button to create
    the bucket.

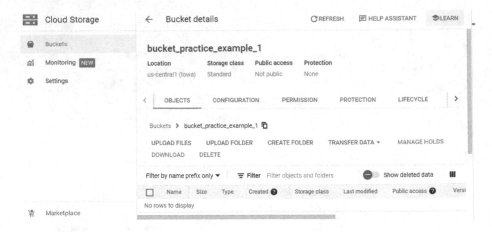

Figure 3-28. *Cloud Storage—new bucket creation*

4.   Once the bucket is created, you can start uploading
     and accessing data in it.

A **custom-trained model** is a machine learning model that
is specifically trained to perform a particular task, such as image
classification, object detection, or natural language processing, based on
custom data.

Custom-trained models are often created to address specific business
needs that pre-trained models cannot meet or to improve the performance
of existing models. They are trained using large amounts of relevant data,
which is used to train the model how to perform the desired task.

In Vertex AI, custom-trained models can be created using a variety
of machine learning frameworks, such as TensorFlow, PyTorch, and
scikit-learn. The process typically involves defining a model architecture,
preparing the training data, and training the model using algorithms such
as gradient descent or backpropagation.

Once a custom-trained model is created, it can be deployed and
integrated into applications to perform the desired task, such as making
predictions or recognizing patterns in new data.

Custom-trained models are an important component of the machine learning workflow in Vertex AI, providing a flexible and powerful solution for solving specific business problems.

Here's how to create a custom-trained model:

1. Project setup.

2. Set up Vertex AI.

3. To create an AutoML machine learning model, turn on the Vertex AI API.

4. Go to the Vertex AI model registry.

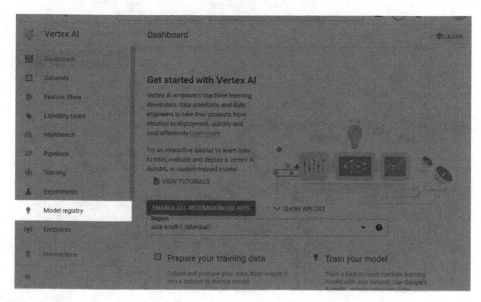

Figure 3-29. *Vertex AI model registry*

5.   Create a custom-trained model.

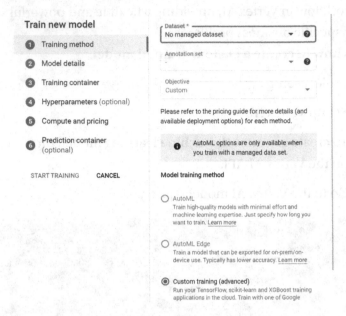

Figure 3-30. *Train a new model*

6.   Click Continue.

Figure 3-31.   *Model details*

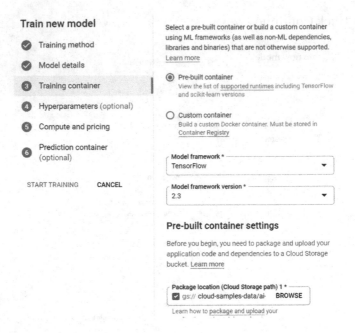

Figure 3-32. *Training container*

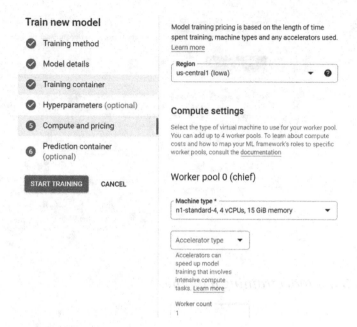

Figure 3-33. Compute and pricing

If everything goes well, after some time the model will appear in the Training menu as the following.

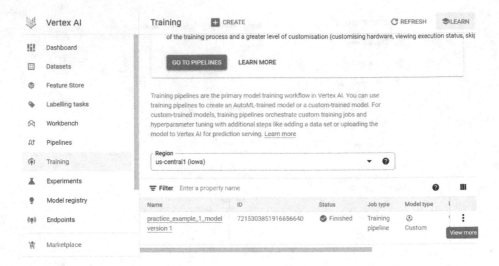

Figure 3-34. *Model training window*

7. To deploy and test a custom-trained model, open Cloud Shell.

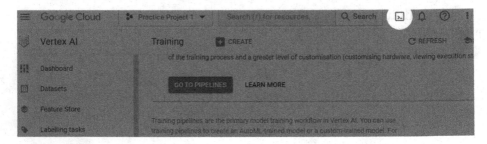

Figure 3-35. *Open Cloud Shell*

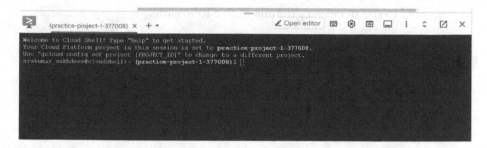

Figure 3-36. *Cloud Shell*

8. Go to Model registry ™ DEPLOY AND TEST.

Figure 3-37. *Deploy and test*

9.   Fill in the necessary details in the following.

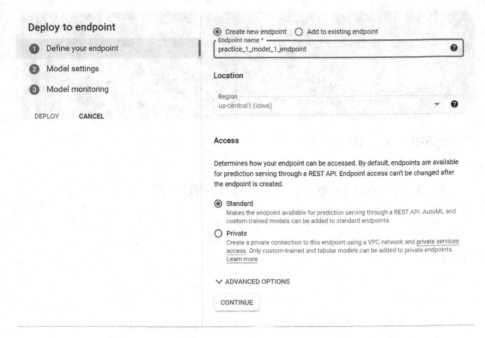

Figure 3-38. *Deploy to endpoint*

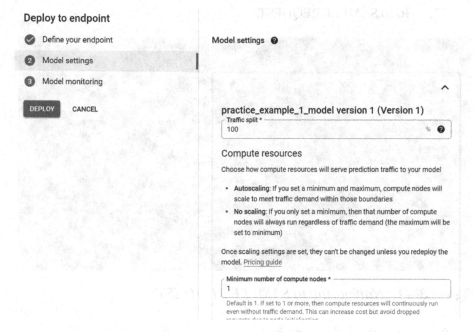

Figure 3-39. *Model settings*

10. Once it's successful, you can see the
     following window.

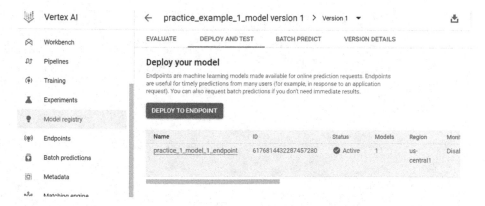

Figure 3-40. *Model deploy to endpoint*

11.   Go to SAMPLE REQUEST.

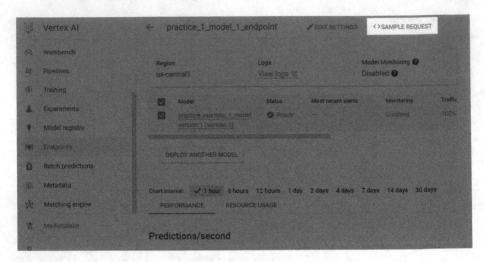

Figure 3-41. *Navigating to SAMPLE REQUEST*

You can see the following window.

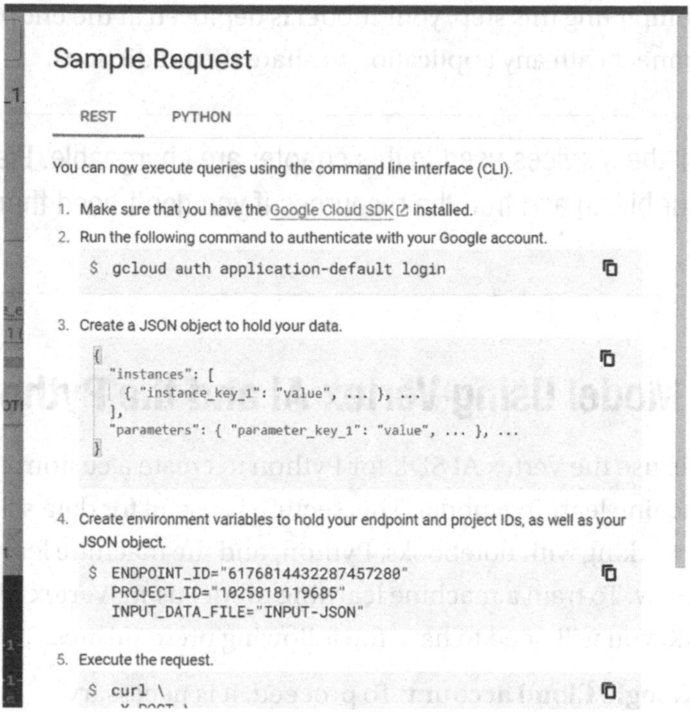

Figure 3-42. *Sample Request*

12. Copy point 4 content from the preceding image by clicking the copy icon and paste in the shell window as the following.

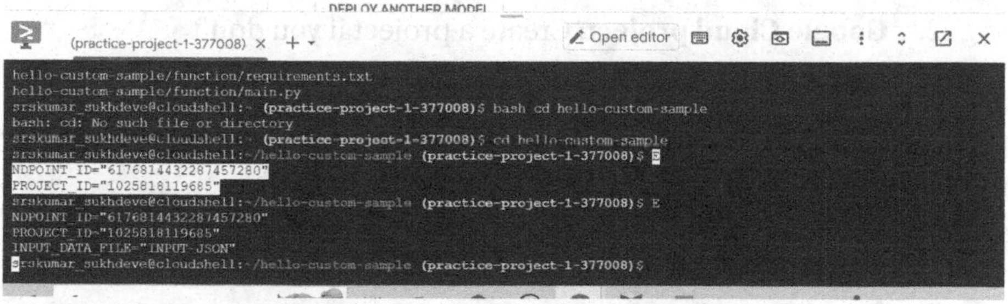

Figure 3-43. *Shell window with the environment variable*

After completing this step, your model is deployed at the endpoint and ready to connect with any application to share the predictions.

Note   All the services used in this chapter are chargeable. Please check your billing and free the resources if you don't need them anymore.

Train a Model Using Vertex AI and the Python SDK

Here we will use the Vertex AI SDK for Python to create a custom-trained machine learning model. This section is useful for data scientists who enjoy working with notebooks, Python, and the machine learning (ML) workflow. To train a machine learning model using Vertex AI and the Python SDK, you will need to have the following prerequisites:

1. **Google Cloud account**: To proceed, it is necessary to have a Google Cloud account. If you do not have one, you can create an account or log in by visiting the following link: `https://console.cloud.google.com/`. If you require guidance on how to create a Google Cloud account, please refer to the earlier chapters of this material.

2. **Google Cloud project**: Create a project if you don't have one using the link `https://console.cloud.google.com/cloud-resource-manager`. Fill in the details as discussed in previous chapters. If you already have a project created, go to the console welcome screen and find the project ID to use later for this experiment.

3. **Vertex AI JupyterLab Notebook**: Follow the steps provided earlier on how to create a user-managed notebook instance.

4. **Open your notebook**: If you are in the Google Cloud Console, search for Vertex AI in the search box.

Figure 3-44. *Search for Vertex AI*

Click Vertex AI. Once you are in Vertex AI, go to **Workbench**.

Figure 3-45. *Open USER-MANAGED NOTEBOOKS*

Click **OPEN JUPYTERLAB**, and you can see the following screen.

Figure 3-46. *JupyterLab*

Now under the notebook, click Python 3. You can rename the notebook as per your choice.

Figure 3-47. *Python notebook*

To train a machine learning model using Vertex AI and the Python SDK, you'll need to perform the following steps:

1. **Install the Vertex AI Python package**: Use the following code in the notebook:

```
# Install the Vertex AI SDK
! pip3 install --upgrade --quiet google-cloud-aiplatform
```

Once you execute the preceding code along with the Vertex AI Python SDK, all its dependencies also get installed. Two of those are Cloud Storage and BigQuery. The --quiet flag is used for suppressing the output so that only errors are displayed if there are any. The exclamation mark (!) is the indication of a shell command.

You can run the code by using Run the selected cells and advance or use the keyboard shortcut "Shift+Enter."

2. **Set the project ID**: The first line of code sets the project ID for the Vertex AI platform. The project ID is a unique identifier for the project in the Google Cloud Console. The project ID is set using the following code:

```
import os

project_id = " MY_PROJECT_ID "
os.environ["GOOGLE_CLOUD_PROJECT"] = project_id
```

Here, the project ID is stored in the project_id variable and set as the active project in the gcloud environment.

3. **Set the region**: The next line of code sets the region for the Vertex AI platform. This is where the Vertex AI resources will be created and stored. The region is set using the following code:

```
region = "us-central1"
```

4. **Create the bucket**: The code then creates a
    new Google Cloud Storage bucket for the Vertex
    AI platform to store its data. A default name is
    provided, but the timestamp is used to make the
    name unique, in case the default name has already
    been taken. The code to create the bucket is as
    follows:

```
bucket_name = "bucket-name-placeholder"  # @param
{type:"string"}
bucket_uri = f"gs://{bucket_name}"

from datetime import datetime
timestamp = datetime.now().strftime("%Y%m%d%H%M%S")

if bucket_name == "" or bucket_name is None or bucket_
name == "bucket-name-placeholder":
    bucket_name = project_id + "aip-" + timestamp
    bucket_uri = "gs://" + bucket_name

from google.cloud import storage
client = storage.Client(project=project_id)

# Create a bucket
bucket = client.create_bucket(bucket_name,
location=region)

print("Bucket {} created.".format(bucket.name))
```

Here, the bucket_name and bucket_uri are defined, and the timestamp
is used to create a unique name for the bucket. Then, a Google Cloud
Storage client is created using the project ID, and the new bucket is created
using the client and the specified name and region.

Initialize the Vertex AI SDK for Python.

The following code initializes the Vertex AI SDK, which provides access to the Vertex AI platform and its capabilities:

```
from google.cloud import aiplatform

# Initialize the Vertex AI SDK
aiplatform.init(project=project_id, location=region, staging_
bucket=bucket_uri)
```

The first line imports the aiplatform module from the google.cloud library, which provides access to the Vertex AI API.

The aiplatform.init method is then called with three arguments: project, location, and staging_bucket. The project argument is the ID of the Google Cloud project from which the Vertex AI platform will be accessed. The location argument is the geographic location where the Vertex AI platform resources will be created. The staging_bucket argument is the URI of the Google Cloud Storage bucket that will be used to store intermediate data during model training and deployment.

This code sets up the connection to the Vertex AI platform and prepares it for use in your application.

Initialize BigQuery.

The following code sets up a BigQuery client to access the BigQuery data storage and analysis service in the specified project:

```
from google.cloud import bigquery

# Set up BigQuery client
bq_client = bigquery.Client(project=project_id)
```

The first line imports the bigquery module from the google.cloud library, which provides access to the BigQuery API.

The bq_client variable is then defined as an instance of the bigquery. Client class, using the project_id variable as an argument. This creates a client object that is used to interact with the BigQuery API for the specified project.

87

The provided code establishes the BigQuery client, enabling access and manipulation of data stored in BigQuery. It grants the ability to execute queries, perform data analysis operations, and undertake various tasks associated with data management within the BigQuery environment.

Create a Vertex AI tabular dataset.

The following code is pre-processing a dataset from BigQuery. It starts by retrieving a table from the BigQuery source dataset (BQ_SOURCE), which is "bigquery-public-data.ml_datasets.penguins". The data is then loaded into a Pandas DataFrame (df).

Next, the code replaces any values in the dataset that match the values defined in the NA_VALUES list ("NA", ".") with NaN and drops any rows with NaN values. This step is done to clean the data and remove unusable rows.

The next step involves converting categorical columns to numeric values. For each of the three categorical columns (island, species, and sex), the code uses the Pandas method factorize to convert the strings in these columns to integers. The factorize method returns a new series and an array of unique values for each column.

The data is then split into two separate data frames: df_train and df_for_prediction. df_train contains 80% of the data and is used for training purposes, while df_for_prediction contains the remaining 20% of the data and is used for predictions.

Finally, the code creates a dictionary for each of the three columns that maps the numeric values back to their string representations. These dictionaries are index_to_island, index_to_species, and index_to_sex. The dictionaries are then printed to verify their values:

```
import numpy as np
import pandas as pd

LABEL_COLUMN = "species"

# Define the BigQuery source dataset
BQ_SOURCE = "bigquery-public-data.ml_datasets.penguins"
```

```
# Define NA values
NA_VALUES = ["NA", "."]

# Download a table
table = bq_client.get_table(BQ_SOURCE)
df = bq_client.list_rows(table).to_dataframe()

# Drop unusable rows
df = df.replace(to_replace=NA_VALUES, value=np.NaN).dropna()

# Convert categorical columns to numeric
df["island"], island_values = pd.factorize(df["island"])
df["species"], species_values = pd.factorize(df["species"])
df["sex"], sex_values = pd.factorize(df["sex"])

# Split into a training and holdout dataset
df_train = df.sample(frac=0.8, random_state=100)
df_for_prediction = df[~df.index.isin(df_train.index)]

# Map numeric values to string values
index_to_island = dict(enumerate(island_values))
index_to_species = dict(enumerate(species_values))
index_to_sex = dict(enumerate(sex_values))

# View the mapped island, species, and sex data
print(index_to_island)
print(index_to_species)
print(index_to_sex)
```

Create a BigQuery dataset.

In the following code snippet

1. The line **bq_dataset_id = f"{project_id}.dataset_id_unique"** creates a string representing the BigQuery dataset ID by concatenating the **project_id** variable and a string "**.dataset_id_unique**".

2. The line **bq_dataset = bigquery.Dataset(bq_ dataset_id)** creates a BigQuery dataset object using the dataset ID string.

3. The line **bq_client.create_dataset(bq_dataset, exists_ok=True)** creates the dataset on BigQuery using the BigQuery client **bq_client**. The **exists_ok** argument is set to **True**, which means that if a dataset with the same ID already exists, the method will not raise an error; instead, it will continue executing the code:

```
# Create a BigQuery dataset
bq_dataset_id = f"{project_id}.dataset_id_unique"
bq_dataset = bigquery.Dataset(bq_dataset_id)
bq_client.create_dataset(bq_dataset, exists_ok=True)
```

Create a Vertex AI tabular dataset

The following code creates a Vertex AI tabular dataset using the **df_ train** Pandas DataFrame. The data in the **df_train** DataFrame will be used to create the tabular dataset. The dataset will be stored in a BigQuery table with the identifier "bq://{bq_dataset_id}.table-unique", and its display name will be "sample-penguins".

The **aiplatform.TabularDataset.create_from_dataframe** method is used to create the tabular dataset from the **df_train** DataFrame. The method takes in three arguments:

- **df_source**: The Pandas DataFrame contains the data to be used to create the tabular dataset.

- **staging_path**: The path to the BigQuery table where the data will be stored.

- **display_name**: The display name for the tabular dataset.

The **bq_dataset_id** variable is used to create the identifier for the BigQuery dataset. The identifier is a string in the format "{project_id}. dataset_id_unique". The **bigquery.Dataset** class is used to create a BigQuery dataset object, and the **bq_client.create_dataset** method is used to create the dataset in BigQuery with the **bq_dataset** object. The **exists_ ok=True** argument is passed to the **bq_client.create_dataset** method to indicate that if the dataset already exists, it should not raise an error, but instead just return the existing dataset:

```
# Create a Vertex AI tabular dataset
dataset = aiplatform.TabularDataset.create_from_dataframe(
    df_source=df_train,
    staging_path=f"bq://{bq_dataset_id}.table-unique",
    display_name="sample-penguins",
)
```

Create a training script.

1. **import argparse** imports the argparse module, which is used to parse command-line arguments passed to the script.

2. **import numpy as np** imports the numpy module and renames it to np for convenience.

3. **import os** imports the os module, which provides a way to interact with the underlying operating system.

4. **import pandas as pd** imports the pandas module and renames it to pd for convenience.

5. **import tensorflow as tf** imports the tensorflow module and renames it to tf for convenience.

6. **from google.cloud import bigquery** imports the bigquery module from the google.cloud package.

7. **from google.cloud import storage** imports the storage module from the google.cloud package.

8. **training_data_uri = os.getenv("AIP_TRAINING_ DATA_URI")** retrieves the value of the AIP_ TRAINING_DATA_URI environmental variable and assigns it to the training_data_uri variable.

9. **validation_data_uri = os.getenv("AIP_ VALIDATION_DATA_URI")** retrieves the value of the AIP_VALIDATION_DATA_URI environmental variable and assigns it to the validation_data_uri variable.

10. **test_data_uri = os.getenv("AIP_TEST_DATA_ URI")** retrieves the value of the AIP_TEST_DATA_ URI environmental variable and assigns it to the test_data_uri variable.

11. **parser = argparse.ArgumentParser()** creates an instance of the argparse.ArgumentParser class.

12. **parser.add_argument('--label_column', required=True, type=str)** adds a required command-line argument --label_column with type str.

13. **parser.add_argument('--epochs', default=10, type=int)** adds a command-line argument --epochs with default value 10 and type int.

14. **parser.add_argument('--batch_size', default=10, type=int)** adds a command-line argument --batch_ size with default value 10 and type int.

15. **args = parser.parse_args()** parses the command-
line arguments and stores them in the args variable.

16. **LABEL_COLUMN = args.label_column** sets the
LABEL_COLUMN variable to the value of the --
label_column argument.

17. **PROJECT_NUMBER = os.environ["CLOUD_ML_
PROJECT_ID"]** retrieves the value of the CLOUD_
ML_PROJECT_ID environmental variable and
assigns it to the PROJECT_NUMBER variable.

18. **bq_client = bigquery.Client(project=PROJECT_
NUMBER)** creates an instance of the bigquery.
Client class with the PROJECT_NUMBER as the
project argument.

```
%%writefile task.py

import argparse
import numpy as np
import os

import pandas as pd
import tensorflow as tf

from google.cloud import bigquery
from google.cloud import storage

# Read environmental variables
training_data_uri = os.getenv("AIP_TRAINING_DATA_URI")
validation_data_uri = os.getenv("AIP_VALIDATION_DATA_URI")
test_data_uri = os.getenv("AIP_TEST_DATA_URI")

# Read args
parser = argparse.ArgumentParser()
```

```
parser.add_argument('--label_column', required=True, type=str)
parser.add_argument('--epochs', default=10, type=int)
parser.add_argument('--batch_size', default=10, type=int)
args = parser.parse_args()

# Set up training variables
LABEL_COLUMN = args.label_column

# See https://cloud.google.com/vertex-ai/docs/workbench/
managed/executor#explicit-project-selection for issues
regarding permissions.
PROJECT_NUMBER = os.environ["CLOUD_ML_PROJECT_ID"]
bq_client = bigquery.Client(project=PROJECT_NUMBER)
```

Next, the following code does the following steps:

1. It defines a function download_table, which takes a BigQuery table URI as an argument.

2. Within the function, the first step is to remove the "bq://" prefix if present in the bq_table_uri.

3. Then, it downloads the BigQuery table as a Pandas DataFrame.

4. After that, it uses the download_table function to download three separate datasets df_train, df_validation, and df_test.

5. Subsequently, a function named "convert_dataframe_to_dataset" is defined, which accepts two DataFrames as parameters. The initial DataFrame corresponds to the training data, while the subsequent DataFrame represents the validation data.

6. Within the function, the first step is to separate the input data and label data. This is done using the. pop method.

7. The input data is then converted to Numpy arrays, and the label data is converted to a one-hot representation using tf.keras.utils.to_categorical.

8. Finally, the data is converted to TensorFlow datasets using tf.data.Dataset.from_tensor_slices for both the training data and validation data.

9. In the end, the function returns the TensorFlow datasets for both the training and validation data.

10. The code then calls the convert_dataframe_to_ dataset function with df_train and df_validation as arguments to create the dataset_train and dataset_ validation datasets.

```
# Download a table
def download_table(bq_table_uri: str):
    # Remove bq:// prefix if present
    prefix = "bq://"
    if bq_table_uri.startswith(prefix):
        bq_table_uri = bq_table_uri[len(prefix) :]

    # Download the BigQuery table as a dataframe
    # This requires the "BigQuery Read Session User" role on
    the custom training service account.
    table = bq_client.get_table(bq_table_uri)
    return bq_client.list_rows(table).to_dataframe()

# Download dataset splits
df_train = download_table(training_data_uri)
df_validation = download_table(validation_data_uri)
```

```
df_test = download_table(test_data_uri)

def convert_dataframe_to_dataset(
    df_train: pd.DataFrame,
    df_validation: pd.DataFrame,
):
    df_train_x, df_train_y = df_train, df_train.
    pop(LABEL_COLUMN)
    df_validation_x, df_validation_y = df_validation, df_
    validation.pop(LABEL_COLUMN)

    y_train = np.asarray(df_train_y).astype("float32")
    y_validation = np.asarray(df_validation_y).
    astype("float32")

    # Convert to numpy representation
    x_train = np.asarray(df_train_x)
    x_test = np.asarray(df_validation_x)

    # Convert to one-hot representation
    num_species = len(df_train_y.unique())
    y_train = tf.keras.utils.to_categorical(y_train, num_
    classes=num_species)
    y_validation = tf.keras.utils.to_categorical(y_validation,
    num_classes=num_species)

    dataset_train = tf.data.Dataset.from_tensor_slices((x_
    train, y_train))
    dataset_validation = tf.data.Dataset.from_tensor_slices((x_
    test, y_validation))
    return (dataset_train, dataset_validation)

# Create datasets
dataset_train, dataset_validation = convert_dataframe_to_
dataset(df_train, df_validation)
```

Next, the following code is a script to train a machine learning model. It consists of the following steps:

1. **Shuffle the train set**: The **dataset_train** is shuffled using the **shuffle** function of the **tf.data.Dataset** class. The argument to **shuffle** is the number of elements in the **df_train** data frame.

2. **Create the model**: The **create_model** function creates the model architecture using TensorFlow's **Sequential** class. The architecture comprises five dense layers with ReLU activation and a softmax output layer. The first layer has 100 neurons, the second layer has 75 neurons, the third layer has 50 neurons, the fourth layer has 25 neurons, and the output layer has three neurons, representing the number of classes. The model is compiled using the **compile** function of the **Sequential** class, with categorical cross-entropy loss, accuracy metric, and RMSprop optimizer.

3. **Batch the datasets**: The **dataset_train** and **dataset_validation** datasets are batched with a specified batch size using the **batch** function of the **tf.data.Dataset** class. The batch size is specified through the **args.batch_size** argument.

4. **Train the model**: The **fit** function of the **Sequential** class is used to train the model on the **dataset_train** data, using the specified number of epochs **args.epochs** and using **dataset_validation** as the validation data.

5. **Save the model**: The trained model is saved to disk using TensorFlow's **tf.saved_model.save** function, with the path to the saved model specified by the **os. getenv("AIP_MODEL_DIR")** environment variable.

```
# Shuffle train set
dataset_train = dataset_train.shuffle(len(df_train))

def create_model(num_features):
    # Create model
    Dense = tf.keras.layers.Dense
    model = tf.keras.Sequential(
        [
            Dense(
                100,
                activation=tf.nn.relu,
                kernel_initializer="uniform",
                input_dim=num_features,
            ),
            Dense(75, activation=tf.nn.relu),
            Dense(50, activation=tf.nn.relu),
            Dense(25, activation=tf.nn.relu),
            Dense(3, activation=tf.nn.softmax),
        ]
    )

    # Compile Keras model
    optimizer = tf.keras.optimizers.RMSprop(lr=0.001)
    model.compile(
        loss="categorical_crossentropy", metrics=["accuracy"],
        optimizer=optimizer
    )
```

```
    return model
```

```
# Create the model
model = create_model(num_features=dataset_train._flat_
shapes[0].dims[0].value)
```

```
# Set up datasets
dataset_train = dataset_train.batch(args.batch_size)
dataset_validation = dataset_validation.batch(args.batch_size)
```

```
# Train the model
model.fit(dataset_train, epochs=args.epochs, validation_
data=dataset_validation)
```

```
tf.saved_model.save(model, os.getenv("AIP_MODEL_DIR"))
```

The output will be **Writing task.py**.

Define arguments for your training script.

The following code sets up the configuration for a machine learning training job:

- The first line, **JOB_NAME**, sets the name of the training job to "custom_job_unique".

- The next two lines, **EPOCHS** and **BATCH_SIZE**, specify the number of epochs and the batch size to be used in the training process.

- The CMDARGS variable is a list that holds the command-line arguments that will be passed to the training script. In this case, there are three command-line arguments being passed:

  1. **"--label_column="** + **LABEL_COLUMN**: Specifies the name of the column that holds the label data. The value for this argument is taken from the LABEL_COLUMN variable.

2. **"--epochs=" + str(EPOCHS)**: Sets the number of epochs to be used in the training process. The value is taken from the EPOCHS variable.

3. **"--batch_size=" + str(BATCH_SIZE)**: Sets the batch size to be used in the training process. The value is taken from the BATCH_SIZE variable.

```
JOB_NAME = "custom_job_unique"

EPOCHS = 20
BATCH_SIZE = 10

CMDARGS = [
    "--label_column=" + LABEL_COLUMN,
    "--epochs=" + str(EPOCHS),
    "--batch_size=" + str(BATCH_SIZE),
]
```

Train and deploy your model.

The following code creates a custom training job on Google Cloud AI Platform—the aiplatform.CustomTrainingJob class is used to specify the details of the training job.

The training job has the following parameters:

- **display_name**: A string that specifies the name of the training job. In this case, the value is "custom_job_unique".

- **script_path**: A string that specifies the path of the script that contains the training code. In this case, the value is "task.py".

- **container_uri**: A string that specifies the Docker container image to use for the training. In this case, the value is "us-docker.pkg.dev/vertex-ai/training/tf-cpu.2-8:latest", which is a TensorFlow CPU image with version 2.8.

- **requirements**: A list of strings that specify the Python packages required for the training. In this case, the list contains two packages: "google-cloud-bigquery>=2.20.0" and "db-dtypes".

- **model_serving_container_image_uri**: A string that specifies the Docker container image to use for serving the model after training. In this case, the value is "us-docker.pkg.dev/vertex-ai/prediction/tf2-cpu.2-8:latest", which is a TensorFlow CPU image with version 2.8 for serving.

The job variable will contain the instance of CustomTrainingJob after the code is executed:

```
job = aiplatform.CustomTrainingJob(
    display_name=JOB_NAME,
    script_path="task.py",
    container_uri="us-docker.pkg.dev/vertex-ai/training/tf-
    cpu.2-8:latest",
    requirements=["google-cloud-bigquery>=2.20.0", "db-
    dtypes"],
    model_serving_container_image_uri="us-docker.pkg.dev/
    vertex-ai/prediction/tf2-cpu.2-8:latest",
)
```

In the following code snippet, the **MODEL_DISPLAY_NAME** variable is defined as "penguins_model_unique". This string is used as the display name for the AI Platform model that will be created during the training process.

The code then calls the **run** method on the **job** object, which represents a custom training job in AI Platform. The **run** method is used to start the training process.

The following parameters are passed to the **run** method:

- **dataset**: A BigQuery dataset object representing the data to be used for training.

- **model_display_name**: A string representing the display name for the AI Platform model that will be created during training. This string is assigned the value of the **MODEL_DISPLAY_NAME** variable.

- **bigquery_destination**: A string representing the destination to store the AI Platform model in BigQuery. The destination is specified as a BigQuery URI, and the **project_id** variable is used to construct the URI.

- **args**: A list of strings representing command-line arguments to be passed to the custom training script. The value of this parameter is assigned the value of the **CMDARGS** variable, which is a list of strings defining the label column to use, the number of epochs to train for, and the batch size.

```
MODEL_DISPLAY_NAME = "penguins_model_unique"

# Start the training and create your model
model = job.run(
    dataset=dataset,
```

```
    model_display_name=MODEL_DISPLAY_NAME,
    bigquery_destination=f"bq://{project_id}",
    args=CMDARGS,
)
```

Wait till you get the output **CustomTrainingJob run completed**.

The following code defines a variable DEPLOYED_NAME, which is assigned the string value "penguins_deployed_unique". This is the display name of the deployed model.

The code then calls the deploy method on the model object, passing in the display name of the deployed model as an argument. The deploy method creates an endpoint to serve predictions based on the trained model. The endpoint is saved with the display name DEPLOYED_NAME and is returned as the endpoint variable:

```
DEPLOYED_NAME = "penguins_deployed_unique"

endpoint = model.deploy(deployed_model_display_
name=DEPLOYED_NAME)
```

Make a prediction.

Prepare prediction test data: The following code removes the "species" column from the data frame df_for_prediction. This is likely because the species column is the label or target column that the model was trained to predict. The species column is not needed for making predictions with the trained model.

Next, the code converts the data frame df_for_prediction to a Python list test_data_list. This is likely to prepare the data for making predictions with the deployed model. The values attribute of the data frame is used to convert the data frame to a Numpy array, and then the tolist() method is used to convert the Numpy array to a list:

```
# Remove the species column
df_for_prediction.pop(LABEL_COLUMN)
```

```
# Convert data to a Python list
test_data_list = df_for_prediction.values.tolist()
```

Send the prediction request: Finally, the code makes predictions by calling the predict() method on the endpoint and passing the test_data_list as the argument for instances. The predictions are stored in the predictions variable and can be accessed by calling predictions.predictions:

```
# Get your predictions.
predictions = endpoint.predict(instances=test_data_list)
```

```
# View the predictions
predictions.predictions
```

Output should look like the following:

```
[[0.323929906, 0.0947248116, 0.58134526],
 [0.300531268, 0.0670630634, 0.632405639],
 [0.349939913, 0.146268025, 0.503792107],
```

The following code uses the argmax method of Numpy to find the best prediction for each set of input data. predictions.predictions is a Numpy array containing the predicted class probabilities for each instance (i.e., each row) of input data. np.argmax(predictions.predictions, axis=1) returns the indices of the highest values along the specified axis (in this case, axis=1 means along each row), effectively giving us the predicted class for each instance. The resulting array, species_predictions, represents the best prediction for the penguin species based on the input data:

```
# Get the prediction for each set of input data.
species_predictions = np.argmax(predictions.
predictions, axis=1)
```

```
# View the best prediction for the penguin characteristics in
each row.
species_predictions
```

Output should look like the following:

```
array([2, 2, 2, 2, 2, 2, 2, 2, 2, 2, 2, 2, 2, 2, 2, 2, 2, 2, 2,
       2, 2, 2,
       2, 2, 2, 2, 2, 2, 2, 2, 2, 2, 2, 2, 2, 2, 2, 2, 2, 2, 2,
       2, 2, 2,
       2, 2, 2, 2, 2, 2, 2, 2, 2, 2, 2, 2, 2, 2, 2, 2, 2, 2, 2,
       2, 2, 2,
       2])
```

Clean up resources.

You can delete the project, which can free all the resources associated with it (you can shut down projects using the Google Cloud Console), or you can use the following steps while retaining the project:

1. Go to the Google Cloud Console.

2. Click the project you used for creating the resources.

3. Click the navigation menu and go to AI Platform.

4. Under AI Platform, select Custom Training Jobs.

5. Click the name of the custom training job that you created and want to delete.

6. On the job details page, click the Delete button. Confirm the deletion when prompted.

7. Go to AI Platform in the navigation menu and select Models.

8. Select the name of the model you wish to remove, and click it.

9. On the model details page, click the Delete button. Confirm the deletion when prompted.

10. Go to BigQuery in the navigation menu.

11. Click the name of the dataset that you created for this project and want to delete.

12. Click the three dots in the upper-right corner and select Delete Dataset. Confirm the deletion when prompted.

For this tutorial, you can also use the following code for deleting the resource:

```
Import os

# Delete the training job
job.delete()

# Delete the endpoint and undeploy the model from it
endpoint.delete(force=True)

# Delete the model
model.delete()

# Delete the storage bucket and its contents
    Bucket.delete(force=True)
```

Note   Make sure to delete all resources associated with the project, as Google Cloud charges for the use of its resources.

Introduction to Google Cloud Dataproc and Its Use Cases for Big Data Processing

Google Cloud Dataproc is a managed cloud service that allows you to easily create and manage Hadoop and Spark clusters. With Dataproc, you can quickly spin up clusters of virtual machines with pre-installed Hadoop, Spark, and other big data frameworks. This capability enables the efficient and cost-effective processing of substantial volumes of data in a scalable manner.

Dataproc integrates with other Google Cloud Platform services, such as Cloud Storage, BigQuery, and Cloud Pub/Sub, making it easy to ingest and export data to and from your clusters. It also includes tools for monitoring and managing your clusters, such as Stackdriver logging and monitoring and Cloud Dataproc Workflow Templates.

One of the key benefits of Dataproc is its flexibility. You have the flexibility to customize the size and configuration of your cluster to align with your specific workload requirements. The payment structure is based on resource usage, ensuring that you only incur costs for the resources you actually utilize.

Dataproc also supports custom images, allowing you to bring your software and configurations to the cluster.

Here are some of the use cases for Google Cloud Dataproc:

1. **ETL processing**: Dataproc can be used to extract, transform, and load (ETL) data from various sources, including log files, social media, and IoT devices.

2. **Machine learning**: Dataproc can be used to train machine learning models on large datasets using popular tools like TensorFlow, PyTorch, and scikit-learn.

3. **Data warehousing**: Dataproc can be used to build data warehouses on top of Google Cloud Storage or other cloud storage services, allowing you to store, analyze, and query large amounts of data.

4. **Real-time data processing**: Dataproc can be used to process streaming data in real time using tools like Apache Kafka, Apache Flink, and Apache Beam.

5. **Business intelligence**: Dataproc offers the capability to construct business intelligence dashboards and generate reports, enabling you to extract valuable insights from data and make well-informed decisions.

Overall, Google Cloud Dataproc is a versatile and scalable tool for processing big data in the cloud, and it can be used in a wide range of industries and applications, including finance, retail, healthcare, and more.

How to Create and Update a Dataproc Cluster by Using the Google Cloud Console

Creating and updating a Dataproc cluster using the Google Cloud Console is a straightforward process. Here are the steps to create and update a Dataproc cluster.

Creating a Dataproc cluster

1. Go to the Google Cloud Console (console.cloud. google.com) and navigate to the Dataproc section.

2. Enable the Dataproc API.

3. Click CREATE CLUSTER to start the cluster creation process.

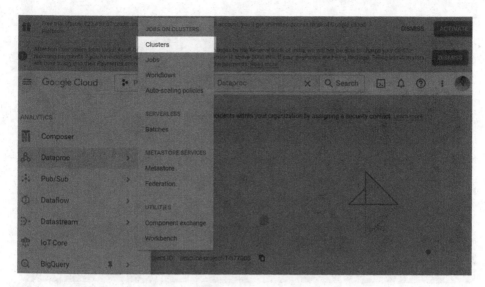

Figure 3-48. *Creating a Dataproc cluster 1*

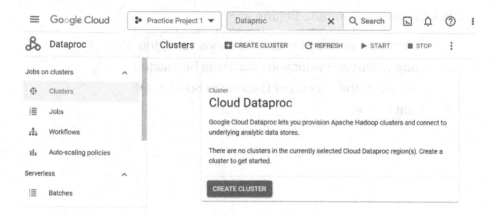

Figure 3-49. *Creating a Dataproc cluster 2*

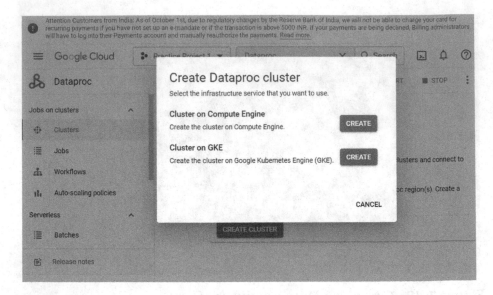

Figure 3-50. *Creating a Dataproc cluster 3*

Click Cluster on Compute Engine.

4. Select a name for your cluster, choose a region and
    zone where you want your cluster to be located,
    and select the version of Hadoop or Spark you
    want to use.

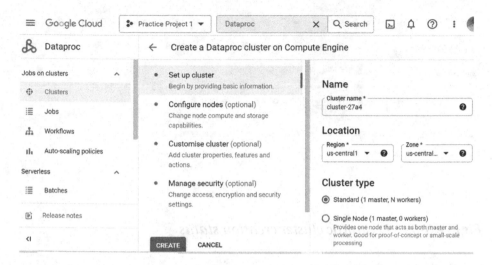

Figure 3-51. *Creating a Dataproc cluster 4*

For proof of concept purposes, you can select the Single Node (1 master, 0 workers) cluster type.

5. Choose the number and type of nodes you want to use for your cluster. You can customize the CPU, RAM, and disk sizes for your nodes.

6. Configure any additional options you want for your cluster, such as initialization scripts or custom metadata.

7. Click Create to start the creation process.

Figure 3-52. *Dataproc cluster creation status*

Updating a Dataproc cluster

1. Go to the Google Cloud Console and navigate to the Dataproc section.

2. Select the cluster you want to update from the list of clusters.

3. Click the EDIT button to start the update process.

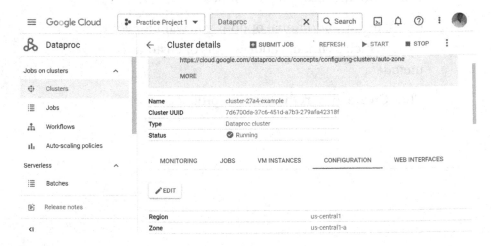

Figure 3-53. *Updating a Dataproc cluster*

4. Modify the settings you want to change, such as
    the number of nodes, node types, or initialization
    actions.

5. Click Save to apply your changes.

Submitting a Spark job

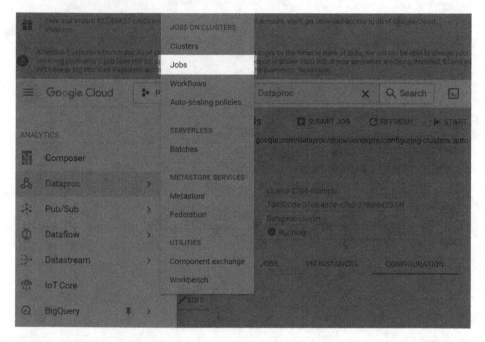

Figure 3-54. *Go to Jobs*

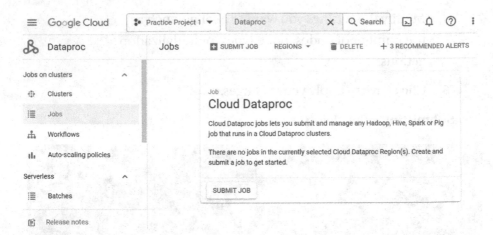

Figure 3-55. *Click SUBMIT JOB*

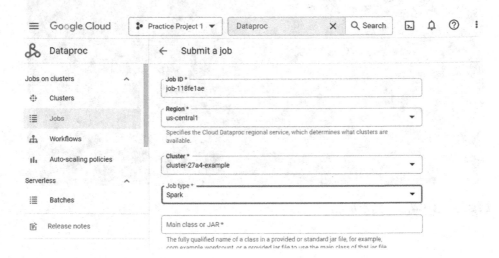

Figure 3-56. *Enter job details*

- **Main class**: org.apache.spark.examples.SparkPi

- **Jar files**: file:///usr/lib/spark/examples/jars/spark-examples.jar

- **Argument**: 10

  Click Submit.

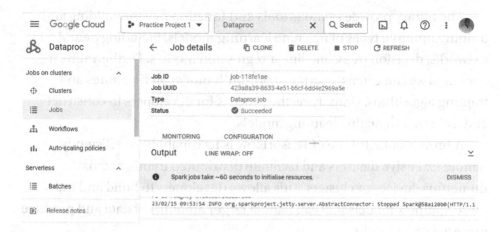

Figure 3-57. Job details

Here in the example, the Spark job uses the Monte Carlo method to estimate the value of Pi. Check for details here: `https://cloud.google.com/architecture/monte-carlo-methods-with-hadoop-spark`.

Note that updating a cluster can cause some downtime or data loss, depending on the type of updates you make. Review the changes carefully and plan accordingly to minimize any impact on your data processing workflows.

Cleaning up

1. On the cluster details page for your created cluster, click Delete to delete the cluster.

2. To confirm deleting the cluster, click Delete.

TensorFlow

TensorFlow is a popular open source software library used for building and training machine learning models. Originating from the Google Brain team in 2015, it has evolved into one of the most extensively employed machine learning libraries globally (Ahmed, 2023).

The library provides various tools and functionality for building and training multiple types of machine learning models, including neural networks, decision trees, and linear regression models. Furthermore, it encompasses an extensive array of pre-built model architectures and training algorithms, simplifying the process for developers to construct and train their machine learning models.

A prominent attribute of TensorFlow is its capability to efficiently handle extensive datasets and facilitate distributed training across numerous devices or clusters. This allows developers to build and train more complex models that can handle larger amounts of data and produce more accurate results.

TensorFlow's versatility and compatibility with various platforms and programming languages, such as Python, C++, and Java, contribute to its noteworthy reputation. This makes it easy to integrate TensorFlow models into a variety of applications and workflows.

In summary, TensorFlow stands as a robust tool for constructing and training machine learning models, encompassing a broad spectrum of features and functionalities that have solidified its status as a favored choice among developers and data scientists.

Here is a simple TensorFlow tutorial that you can run on Colab.

In this tutorial, we will build a simple TensorFlow model to classify images from the MNIST dataset.

Step 1: Set up the environment.

The first step is to set up the environment on Colab. We will install TensorFlow 2.0 and import the necessary libraries:

```
!pip install tensorflow==2.0.0-alpha0
```

```
import tensorflow as tf
from tensorflow import keras
```

Step 2: Load and preprocess data.

Next, we will load and preprocess the data. We will use the MNIST dataset, which contains 60,000 training images and 10,000 test images of handwritten digits (Reis, 2016). The following code loads the MNIST dataset and preprocesses it by scaling the pixel values between 0 and 1 and converting the labels to one-hot encoding:

```
(x_train, y_train), (x_test, y_test) = keras.datasets.mnist.
load_data()

x_train = x_train.astype('float32') / 255.0
x_test = x_test.astype('float32') / 255.0

y_train = keras.utils.to_categorical(y_train)
y_test = keras.utils.to_categorical(y_test)
```

Step 3: Build the model.

Now, we will build our model using TensorFlow's Keras API. We will use a simple neural network with two hidden layers. The following code defines a simple neural network with two hidden layers using the Keras API. The model consists of a Flatten layer that flattens the input image into a vector, followed by two dense layers with ReLU activation, and a final dense layer with softmax activation that produces the output probabilities:

```
model = keras.Sequential([
    keras.layers.Flatten(input_shape=(28, 28)),
    keras.layers.Dense(128, activation='relu'),
    keras.layers.Dense(10, activation='softmax')
])
```

Step 4: Train the model.

Next, we will train the model on the training data using the **compile** and **fit** methods. The following code compiles the model with the Adam optimizer and categorical cross-entropy loss function and trains it on the training data for five epochs with a batch size of 32:

```
model.compile(optimizer='adam',
              loss='categorical_crossentropy',
              metrics=['accuracy'])

model.fit(x_train, y_train, epochs=5, batch_size=32)
```

Step 5: Evaluate the model.

Finally, we will evaluate the model on the test data and print the accuracy:

```
test_loss, test_acc = model.evaluate(x_test, y_test)

print('Test accuracy:', test_acc)
```

That's it! You have successfully built and trained a TensorFlow model on Colab.

To run this tutorial on Colab, simply create a new notebook and copy-paste the code into it. Make sure to select a GPU runtime for faster training.

Summary

In this chapter, we delved into the world of big data and machine learning, exploring several Google Cloud services and tools.

We started with an introduction to BigQuery, a fully managed data warehouse solution that enables storing and querying massive datasets. We discussed its use cases and how it facilitates large-scale data analytics. We have implemented it using Sandbox, which is free to use.

Next, we explored BigQuery ML, a feature of BigQuery that allows us to build machine learning models directly within the BigQuery environment. We learned how to create and deploy models using SQL queries, simplifying the machine learning process.

Moving on, we were introduced to Google Cloud AI Platform, a comprehensive platform offering a range of machine learning tools and services. We discovered its capabilities for training and deploying machine learning models, providing us with a scalable and flexible infrastructure.

We then focused on Vertex AI, a powerful machine learning platform that offers a unified interface for training and deploying models. We discussed the benefits of using Vertex AI, such as effectively managing and scaling machine learning workflows.

Additionally, we explored Google Cloud Dataproc, a managed Spark and Hadoop service designed for big data processing. We learned about its use cases and how it simplifies the deployment and management of big data processing clusters.

Finally, we delved into TensorFlow, a popular open source machine learning framework. We gained an understanding of its basics and its capabilities for building and training machine learning models.

Throughout the chapter, we learned about the different tools and services available on Google Cloud for working with big data and machine learning. This knowledge equips us with the necessary skills to effectively leverage these tools in our big data and machine learning projects.

CHAPTER 4

Data Visualization and Business Intelligence

Within this chapter, our focus will be directed toward the realm of data visualization and its intersection with business intelligence. Our attention will be particularly drawn to Looker Studio and Colab as potent instruments for crafting engaging visualizations and comprehensive reports.

Embarking on our journey, we shall commence with an exploration of the capabilities of Looker Studio, a robust tool for both data investigation and visualization. Through this, we will unravel its array of features that empower us to convert raw data into meaningful insights.

Subsequently, we will plunge into the intricacies of producing and disseminating data visualizations and reports via Looker Studio. This educational segment will provide insights into establishing connections with diverse data sources, architecting visualizations, and fabricating interactive reports, easily shared with concerned parties.

Our exploration will then shift to the fusion of BigQuery and Looker, illuminating how these two tools harmoniously collaborate to dissect and visualize extensive datasets. A comprehension of the synergy between BigQuery's potent querying prowess and Looker's user-friendly interface will be cultivated.

© Shitalkumar R. Sukhdeve and Sandika S. Sukhdeve 2023
S. R. Sukhdeve and S. S. Sukhdeve, *Google Cloud Platform for Data Science*,
https://doi.org/10.1007/978-1-4842-9688-2_4

Advancing, we will home in on the art of constructing dashboards using Looker Studio. This phase will reveal the techniques behind crafting tailored dashboards that coalesce numerous visualizations, offering a panoramic snapshot of vital metrics and performance indicators.

Concluding our expedition, we will delve into data visualization through Colab, a collaborative notebook milieu. Through this segment, we will assimilate the adept utilization of Python libraries such as pygwalker, panda and numpy to yield visually captivating charts and graphs within the Colab framework.

Throughout this chapter's progression, the importance of data visualization will be illuminated, particularly its pivotal role in steering business intelligence. We will adeptly harness Looker Studio and allied tools to transmute intricate data into actionable insights, effectively communicating them to stakeholders in a captivating and comprehensible manner.

Looker Studio and Its Features

Google Looker Studio is a data analytics and business intelligence platform that is integrated with Google Cloud. It provides a unified interface for exploring, visualizing, and sharing data insights with others.

Key features of Looker Studio are as follows:

1. **Data modeling**: Looker Studio allows you to build a data model by defining the relationships between your data tables and creating custom calculations.

2. **Interactive data visualizations**: The platform provides a wide range of interactive data visualizations, including bar charts, line graphs, pie charts, and heat maps.

3. **Dashboards and reports**: Looker Studio allows you to create interactive dashboards and reports to share data insights with stakeholders.

4. **SQL editor**: The SQL editor provides a user-friendly interface for writing and executing SQL queries, making it easier to extract data from databases and manipulate it.

5. **Collaboration and sharing**: Looker Studio has robust collaboration and sharing features, allowing you to share dashboards and reports with others and make real-time changes.

6. **Integrations**: Looker Studio integrates with various data sources, including Google BigQuery, Amazon Redshift, Snowflake, and more.

7. **Customizable LookML**: Looker Studio defines data models using LookML, a proprietary modeling language. LookML is highly customizable, allowing you to create custom calculations and build your data models.

8. **Advanced security**: To safeguard your data, Looker Studio offers a suite of advanced security features, including data encryption, access control, and user management. These measures work in unison to ensure the protection and integrity of your valuable data.

These are some of the key features of Google Looker Studio. With its powerful data modeling, interactive data visualizations, and collaboration features, Looker Studio makes exploring, understanding, and sharing data insights with others easier.

Creating and Sharing Data Visualizations and Reports with Looker Studio

Here's a step-by-step tutorial for using Looker Studio.

Sign up for a Looker account: if you don't already have a Looker account, sign up for one on the Looker website.

1. **Log into Looker**: Once you have an account, log into your Looker account using the link `https://lookerstudio.google.com/navigation/reporting`. You can see the following screen.

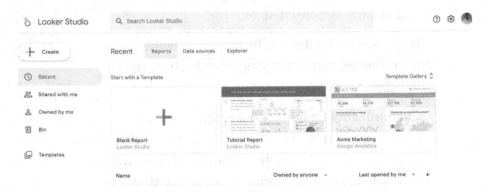

Figure 4-1. *Welcome screen of Looker Studio*

Click Create and select Report.

2. **Connect to your data**: Looker Studio needs access to your data to build reports and visualizations. Connect your data sources to Looker by going to the Data tab and clicking Connections. Follow the instructions to add your data sources.

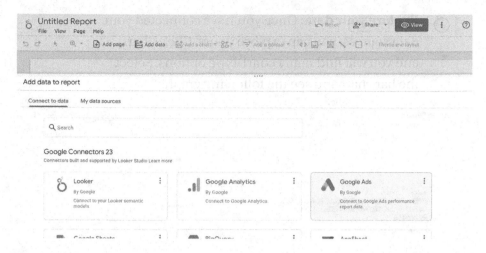

Figure 4-2. *New report*

For this example, we can go to My data sources and select [Sample] World Population Data 2005 - 2014, which is freely available. Click Add to report. The following screen will appear.

Figure 4-3. *Data imported to the report*

3. **Explore your data**: Once you have connected your data sources, you can explore yours. Click Chart, and you will find many chart-type options. Select the bar chart and see the following graph.

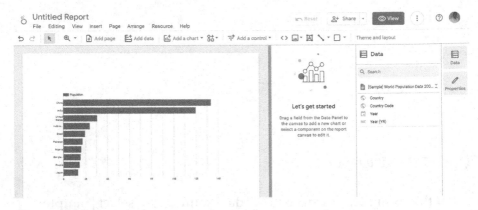

Figure 4-4. *Visualization from data*

4. **Create a Look**: You can see the final output by clicking the View button at the top right of the preceding screen.

5. **Save and share your work**: Once you have built your visualizations, you can save and share your work. Go to the Share tab and select how you want to share your work. Sharing your data visualization or report becomes effortless through various means, including sharing a link, embedding it on a website, or exporting it to a different format. These options provide flexibility in disseminating your work to others.

6. **Collaborate with others**: Looker Studio incorporates collaborative functionalities that enable seamless teamwork on reports and visualizations. You can share your work with others, granting them permission to either edit or view your content, fostering effective collaboration among team members.

Here's how to add data from Excel and explore:

1. **Prepare your file:** The Excel file should be in a proper format. See the following example. The heading should start with the first-row first column, with no vacant space above it.

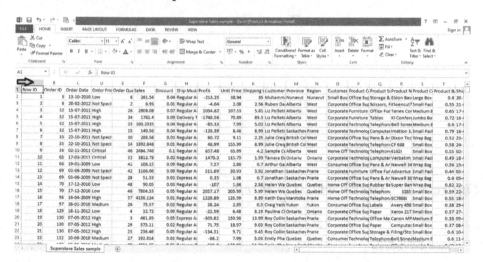

Figure 4-5. *Sample Excel file*

2.   Select the entire header row and save it as a CSV file.

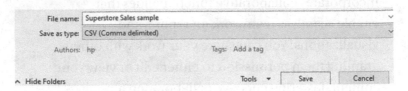

Figure 4-6. *Save the file in CSV format*

3.   Go to Looker.

4.   Create a new report by clicking Create.

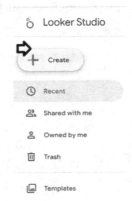

Figure 4-7. *Create a new Looker report*

5. Go to Data source.

Figure 4-8. *Data source tab*

6. Scroll to File Upload and click it.

◊ Untitled Data Source

Q Search

Connect to Google Sheets.	Connect to BigQuery tables and custom queries.	Connect to AppSheet app data.

⬆ File Upload ⋮	🔴 Amazon Redshift ⋮	🔗 Campaign Manager 360 ⋮
By Google	By Google	By Google
Connect to CSV (comma-separated values) files.	Connect to Amazon Redshift.	Connect to Campaign Manager 360 data.

▲ Cloud Spanner ⋮	◉ Cloud SQL for MySQL ⋮	▶ Display & Video 360 ⋮

Figure 4-9. *Select File Upload*

7.   Upload the file.

Figure 4-10.   *Upload data*

8.   Rename the file, currently titled Data Source.

Figure 4-11.   *Rename the title of Data Source*

9.   The file should show ● Uploaded.

Superstore Sales report.csv

TOTAL FILE SIZE	NUMBER OF FILES	CREATION DATE	LAST MODIFIED DATE
2 MB (2% of 100MB used)	1	2/14/23 10:41 AM	2/14/23 10:42 AM

VIEW FILES IN CLOUD

ADD FILES   Files must contain the same schema. Learn More

File name	Uploaded at	Size	Status
Superstore Sales report.csv	2/14/23 10:42 AM	2 MB	● Uploaded

DELETE DATA SET

Figure 4-12.   *Data upload status*

10. Connect the file with the CONNECT button placed on the top right-hand side.

CONNECT

lmost any source by uploading CSV (comma-separated values) files. File upload lets
c connector.

Superstore Sales report.csv

TOTAL FILE SIZE	NUMBER OF FILES	CREATION DATE	LAST MODIFIED DATE
2 MB (2% of 100MB used)	1	2/14/23 10:41 AM	2/14/23 10:42 AM

VIEW FILES IN CLOUD

ADD FILES   Files must contain the same schema. Learn More

File name	Uploaded at	Size	Status
Superstore Sales report.csv	2/14/23 10:42 AM	2 MB	● Uploaded

DELETE DATA SET

Figure 4-13. *Connect the data*

11. After the file gets connected, you will see the view like the following screenshot. Then check the exact format of the dimensions. For example, Customer Name is in Text type, Order Date is in Date type, Order Quantity is in Number type, etc.

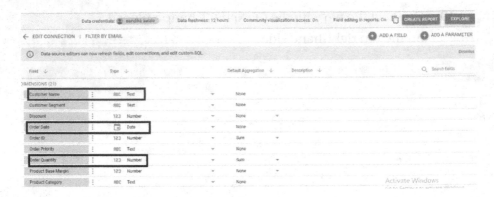

Figure 4-14. *Data view*

12. Then click CREATE REPORT. It will move you to a new view. You will see the following screen.

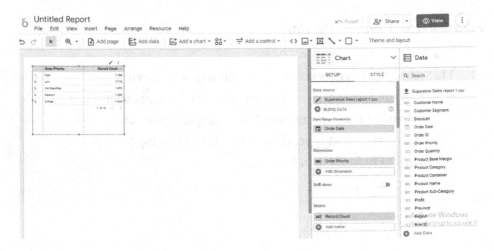

Figure 4-15. *First appearance after creating a report*

13.   Name the untitled report file.

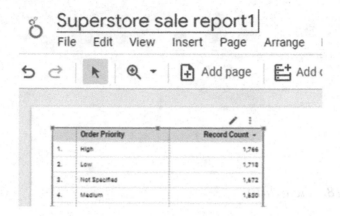

Figure 4-16. *Give the report a name*

This is how you can create a report and visualize your data.

14.   You can see the table in view mode.

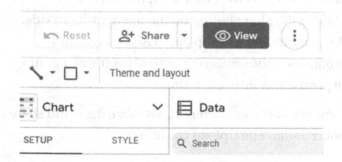

Figure 4-17. *View mode*

It looks like this one.

Figure 4-18. *View data*

You can edit the file by clicking the Edit button on the top right-hand side. It will again go back to its original form before view mode.

15.   Share the report with your team members by clicking the Share button on the top right-hand side. A share window will pop up. In Add people and groups, add the email ID/name of people or a group of people and save. Click Done.

This is how you can create a report, visualize data, and share the report with the Looker Studio File Upload connector.

BigQuery and Looker

As we have already gone through BigQuery learning and created some tables, we can use the tables in Looker for visualization.

1.   Go to Report in Looker and click Add data.

2.   Go to Connect to data and select BigQuery and allow authorization.

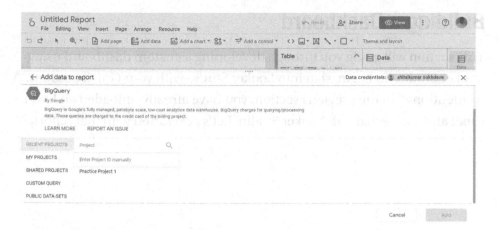

Figure 4-19. *BigQuery and Looker*

You should be able to view the screen depicted here. As we have already created the project in previous chapters, we can use the same, or if you have not, please follow the last chapters.

3. Select the project shown in the preceding screen, Practice Project 1.

4. You will see a list of datasets available with your project. If you are using Looker, which is freely available, and a dataset from BigQuery Sandbox, it won't be counted in billing (please check the latest Google guidelines for the same for pricing).

5. Click a dataset and select a table. You will be able to see your data in the report.

Building a Dashboard

This section will guide you through developing a custom dashboard using Looker Studio. To begin, sign into Looker Studio with your Gmail account. As mentioned in the previous section, you have already uploaded an Excel sheet and connected it to Looker Studio. Let's get started with that report.

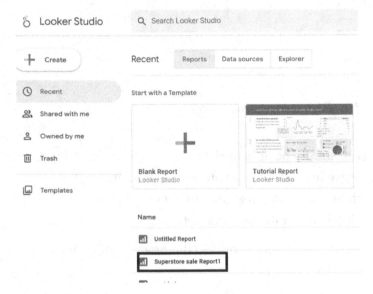

Figure 4-20. *Select a report*

Return to your previous report and click the Page option in the header. From there, select New page to create a blank report for you to work with.

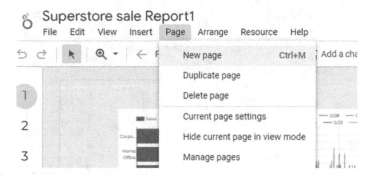

Figure 4-21. *Insert a new page*

Change the theme and layout as per your choice.

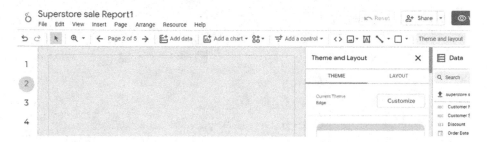

Figure 4-22. *Select Theme and Layout*

Now, we can start adding different charts to build the dashboard:

1. **Scorecard:** Go to the Add a chart button and click
   the scorecard Total. Create multiple scorecards by
   adding more scorecards or by copying and pasting
   the given scorecard. Change the metric of each
   scorecard as shown in the Setup section on the
   right-hand side of the page.

Figure 4-23. *Create scoreboards*

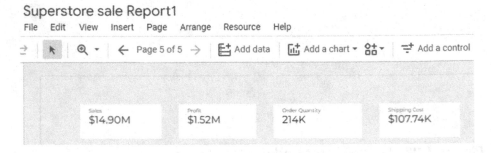

Figure 4-24. *Scoreboards for building a dashboard*

2. **Time series chart**: Click the Add a chart button and
   select Time series from the options. Drag the chart
   to appear below the Sales scorecard. Customize
   the chart to display the sales trend during a specific
   period. To do this, go to the Setup option, select
   Default date range, and choose the Custom option.
   Then, select the start and end dates for the sales
   trend. See the following example.

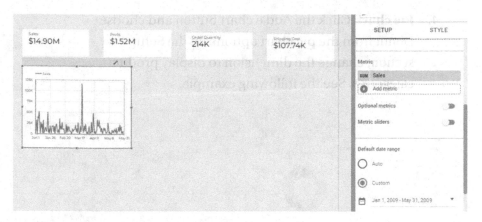

Figure 4-25. *Line chart*

3. **Bar chart**: Click the Add a chart button and choose
    Bar from the options. Drag the chart onto the page
    concerning the previous time series chart. Add
    another metric, such as the profit, to the chart. See
    the following example.

Figure 4-26. *Bar chart*

4. **Pie chart**: Click the Add a chart button and choose Donut from the pie chart options. In the Setup section, change the dimension to display product categories. See the following example.

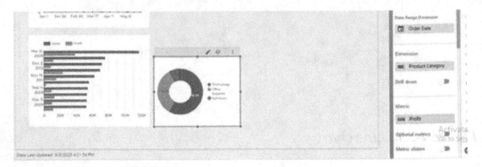

Figure 4-27. *Donut chart*

5. **Table**: Choose the Table option from the Add a chart button. Create a table that displays the sales and profit of each product category.

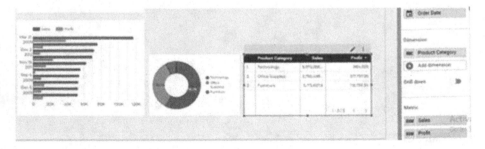

Figure 4-28. *Table*

6. **Bubble map**: Click the Add a chart button and choose Bubble Map from the Google Maps options. Position the map with the line chart on the page by dragging it. Adjust the setup settings by selecting Location and choosing Region, setting Tooltip to display Province, selecting Region as Color dimension, and using Sales or Profit for Size.

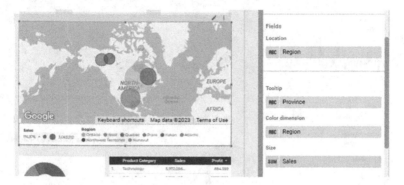

Figure 4-29. *Bubble map*

7. **Date range**: Click the Add a control option and select Date range control. Drag the control to the top-right corner of the page. Select the start and end dates for the range, which will adjust all the chart values on the dashboard accordingly.

Figure 4-30. *Date range control*

The completed dashboard is presented in the following.

Figure 4-31. *Dashboard*

Data Visualization on Colab

PyGWalker is a Python library that simplifies the data analysis and data visualization workflow in Jupyter Notebook. It provides a Tableau-style user interface for the visual exploration of data frames in Pandas. With PyGWalker, data scientists can quickly and easily analyze data and visualize patterns with simple drag-and-drop operations.

PyGWalker is built on top of the open source software Graphic Walker, which is an alternative to Tableau. Graphic Walker provides a visual interface that allows users to explore real-time data. It also supports various chart types and data manipulation functions.

To get started with PyGWalker, you need to install it using pip. Open a Jupyter Notebook and install the library as follows:

```
!pip install pygwalker
```

Next, we can import PyGWalker and our other required libraries:

```
import pandas as pd
import numpy as np
import pygwalker as pyg
```

For this tutorial, we will create a sample dataset with some random data:

```
data = {
    'age': np.random.randint(18, 65, 100),
    'gender': np.random.choice(['Male', 'Female'], 100),
    'income': np.random.normal(50000, 10000, 100),
    'state': np.random.choice(['CA', 'NY', 'TX'], 100)
}

df = pd.DataFrame(data)
```

Now, let's use PyGWalker to visualize this dataset:

```
pyg.walk(df)
```

You can see the following output.

Figure 4-32. *PyGWalker Data tab*

PyGWalker will automatically detect the data types and create a Tableau-style interface.

Figure 4-33. *PyGWalker Visualization tab*

We can use drag-and-drop operations to explore and visualize our data. For example, we can drag the age column to the x-axis and the income column to the y-axis to create a scatter plot. Before doing that, first, click age, and the bin suggestion will appear, so click that. Do the same with income. Now drag and drop bin(age) and bin(income) onto the x- and y-axis, respectively. You can find the following output.

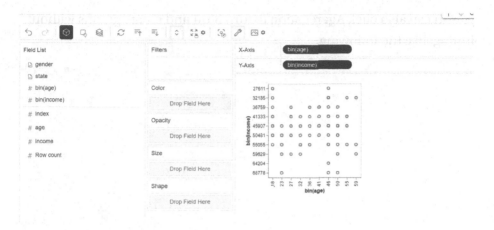

Figure 4-34. *Scatter plot*

We can also add additional dimensions to our visualization by dragging other columns to the color or size shelf. For example, we can drag the gender column to the color shelf to create a scatter plot with different colors for each gender.

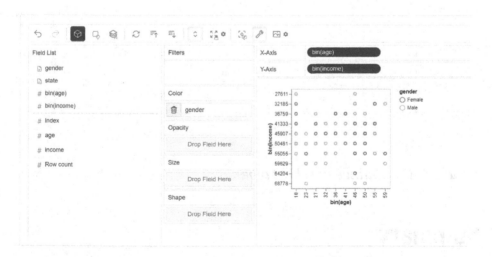

Figure 4-35. *Adding another dimension to the scatter plot*

145

You can also click Aggregation to turn it off and create graphs without binning, like the following.

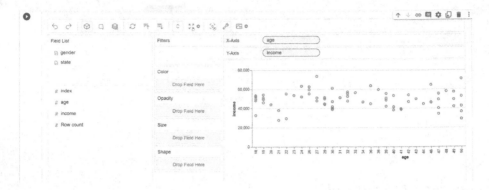

Figure 4-36. *Without aggregation*

You can also change the chart type like the following.

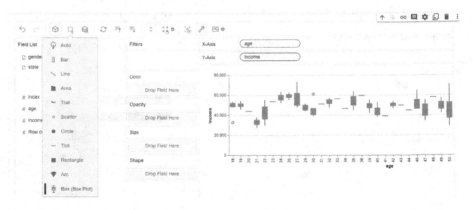

Figure 4-37. *Change the chart type*

Summary

In this chapter, we explored the world of data visualization and business intelligence, focusing on Looker Studio and Colab for creating compelling visualizations and reports.

We began with an introduction to Looker Studio, a powerful data exploration and visualization tool. We discovered its features and capabilities for transforming raw data into meaningful insights.

Next, we delved into the process of creating and sharing data visualizations and reports using Looker Studio. We learned how to connect to various data sources, design visualizations, and generate interactive reports that can be easily shared with stakeholders.

We then explored the integration between BigQuery and Looker, understanding how these two tools can be used together to analyze and visualize large datasets. We learned about the benefits of leveraging BigQuery's powerful querying capabilities and Looker's intuitive interface for data exploration.

Moving on, we focused on building dashboards using Looker Studio. We discovered how to create customized dashboards that consolidate multiple visualizations and provide a comprehensive view of key metrics and performance indicators.

Lastly, we explored data visualization with Colab, a collaborative notebook environment. We learned to leverage Python libraries like pygwalker, panda and numpy to create visually appealing charts and graphs within a Colab notebook.

Throughout the chapter, we gained insights into the importance of data visualization and its role in driving business intelligence. We learned how to effectively use Looker Studio and other tools to transform complex data into actionable insights and communicate them to stakeholders in a compelling and accessible manner. This knowledge equips us with the skills to make informed decisions and drive business success through data visualization and business intelligence.

CHAPTER 5

Data Processing and Transformation

In this chapter, we will explore the domain of data manipulation and alteration, centering our attention on Google Cloud Dataflow and Google Cloud Dataprep.

Our journey begins with an initiation into Google Cloud Dataflow, an influential service dedicated to both batch and stream data handling. We will delve into the diverse scenarios where Dataflow stands out, encompassing real-time analytics, ETL (Extract, Transform, Load) pipelines, and event-triggered processing. This will foster an understanding of Dataflow's role in facilitating scalable and efficient data manipulation.

Progressing onward, we will immerse ourselves in the execution of data manipulation pipelines within Cloud Dataflow. This phase involves comprehending the concept of pipelines, a framework that enables the definition of a sequence of steps required to process and reshape data. We will investigate how Dataflow proficiently accommodates batch and stream data manipulation and delve into its seamless integration with other Google Cloud services, notably BigQuery and Cloud Pub/Sub.

Steering ahead, our exploration takes us to Google Cloud Dataprep, a service tailored for data refinement and metamorphosis. We will explore a spectrum of contexts where Dataprep proves its utility, including data scrubbing, normalization, and data refinement. The user-friendly visual interface of Dataprep will be unveiled, showcasing its ability to facilitate effortless data exploration and manipulation, bypassing the need for intricate coding.

The entirety of this chapter will grant us insights into the pivotal role that data manipulation and alteration assume in the data lifecycle. Through the lens of Google Cloud Dataflow and Google Cloud Dataprep, we will grasp the potent tools at our disposal for processing and refining data on a substantial scale. This proficiency will enable us to unearth valuable insights and steer well-informed decision-making.

By harnessing the capabilities of Dataflow and Dataprep, we will cultivate expertise and adeptness in devising and executing streamlined data manipulation pipelines. This adeptness ensures the conversion of raw data into a usable format, ensuring the integrity and uniformity of data. This empowerment enables the full exploration of our data's potential, extracting invaluable insights that catalyze business expansion and triumph.

Introduction to Google Cloud Dataflow and Its Use Cases for Batch and Stream Data Processing

Google Cloud Dataflow is a comprehensive managed cloud service that facilitates the development and execution of diverse data processing patterns, encompassing both batch and stream processing. It streamlines the process by offering a unified programming model, enabling users to concentrate on crafting business logic rather than dealing with the intricacies of infrastructure management.

Some common use cases for batch processing with Google Cloud Dataflow include data warehousing, ETL (Extract, Transform, Load) processing, and analytics. For example, you can use Dataflow to extract data from multiple sources, clean and transform the data, and load it into a data warehousing solution like BigQuery for analysis.

For stream processing, Dataflow is used for real-time analytics, fraud detection, IoT data processing, and clickstream analysis. For example, you can use Dataflow to process real-time data from IoT devices and analyze the data to gain insights and make decisions in real time.

Overall, Google Cloud Dataflow is a highly scalable and flexible data processing solution that can handle both batch and stream processing workloads, making it a valuable tool for data engineers and data scientists.

Running Data Processing Pipelines on Cloud Dataflow

There are several ways to utilize Dataflow for your data processing needs. You can create Dataflow jobs using either the cloud console UI, the gcloud CLI, or the API, all of which offer various options for job creation. Additionally, Dataflow templates provide a selection of pre-built templates for easy use, and you can also create custom templates and share them with colleagues.

With Dataflow SQL, you can leverage your SQL skills to develop streaming pipelines directly from the BigQuery web UI, enabling you to join data from various sources, write results into BigQuery, and visualize them in real-time dashboards. Moreover, the Dataflow interface provides access to Vertex AI Notebooks, which allow you to build and deploy data pipelines using the latest data science and machine learning frameworks (Vergadia, 2021). Finally, Dataflow inline monitoring allows you to access job metrics directly, providing helpful insights for troubleshooting pipelines at both the step and worker levels.

Introduction to Google Cloud Dataprep and Its Use Cases for Data Preparation

Google Cloud Dataprep is an all-inclusive data preparation service managed by Google. It empowers users to visually explore, cleanse, and prepare their data for analysis or machine learning purposes. With its intuitive interface, users can interactively delve into and modify their data without the need for coding, simplifying the data preparation process.

Here are some of the use cases of Google Cloud Dataprep for data preparation:

Data cleansing: Cloud Dataprep provides a comprehensive set of tools to clean and standardize your data, ensuring consistency, accuracy, and completeness. It offers features like deduplication, text parsing, regular expressions, data formatting, and normalization. These capabilities enable you to effectively process and enhance your data for improved analysis and reliability.

Data transformation: Cloud Dataprep lets you transform your data into the desired format for downstream analysis or machine learning. It includes features such as pivoting, aggregating, and merging datasets.

Data enrichment: You can use Cloud Dataprep to enrich your data by joining it with external data sources such as databases or APIs. This can help you gain deeper insights into your data and create more comprehensive analyses.

Data exploration: Cloud Dataprep lets you quickly explore your data using its visual interface. You can use the data profiling feature to understand the distribution of your data and identify outliers, anomalies, or missing values.

Collaboration: Cloud Dataprep provides a collaborative environment for teams to work on data preparation projects. It enables multiple users to simultaneously work on the same project, making sharing ideas, data, and insights easier.

In summary, Google Cloud Dataprep is a powerful tool for data preparation that can help you streamline your data preparation workflows, reduce the time and effort required to prepare your data, and improve the accuracy and quality of your analyses.

1. Enable Dataflow and required APIs in a Google Cloud project:

   - Dataflow API

   - Compute Engine API

   - Cloud Logging API

   - Cloud Storage

   - Google Cloud Storage JSON API

   - BigQuery API

   - Cloud Pub/Sub API

   - Cloud Resource Manager API

2. Create a Cloud Storage bucket:

    - Log into the Google Cloud Console at `https://console.cloud.google.com/`.

    - Select your project, or create a new one if you haven't already.

    - Open the Cloud Storage page from the main menu.

    - Click the Create bucket button.

    - In the Create a bucket dialog box, enter a unique name for your bucket.

    - Select the region where you want your data to be stored.

    - Choose the default storage class for your bucket. You can also choose a different storage class based on your needs.

    - Choose your bucket's access control settings. You can choose to keep your bucket private or make it publicly accessible.

    - Click Create to create your new bucket.

3. Create a BigQuery dataset.

4. Create a BigQuery table:

    - In the navigation pane, expand your project and select the dataset where you want to create the new table.

    - Click the Create table button.

    - In the Create table dialog box, enter a name for your new table.

- Choose the source of your data. You can create a new table from scratch or use an existing one as a template.

- In the Schema section, define the columns for your table by specifying the name, data type, and mode for each field.

- If your data source is a CSV or JSON file, specify the file format and the location of the file.

- Configure the advanced options for your table, such as partitioning and clustering.

- Click Create table to create your new BigQuery table.

5. Run the pipeline.

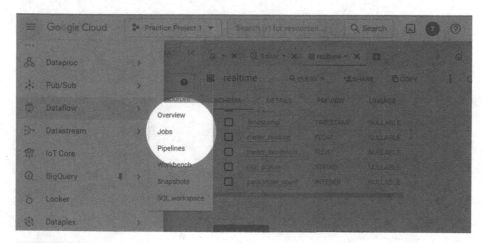

Figure 5-1. *Dataflow*

6.   Click CREATE JOB FROM TEMPLATE.

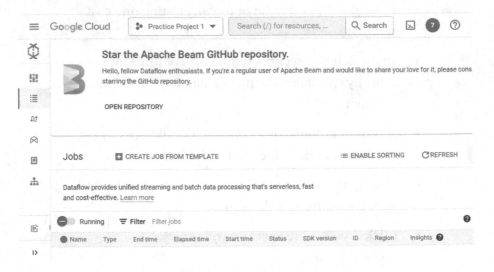

Figure 5-2.   *Star the Apache Beam GithHub repository*

7.   Run the pipeline.

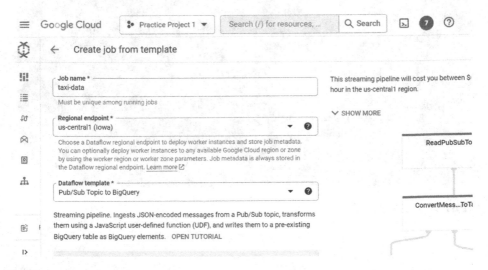

Figure 5-3.   *Create a job from a template*

8. After running the pipeline, you can see the
    following screen.

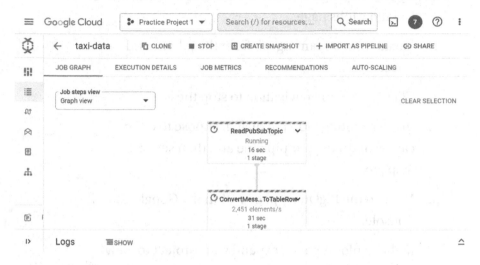

Figure 5-4. *After pipeline execution*

9. View your results.

Figure 5-5. *View the data written to the real-time table*

You have effectively set up a streaming process that retrieves JSON-formatted messages from a Pub/Sub topic and transfers them to a BigQuery table.

Clean up:

1. Open the Google Cloud Console and navigate to the Dataflow Jobs page.

2. Locate your streaming job from the job list and select it.

3. Click Stop in the navigation to stop the job.

4. In the resulting Stop job dialog, choose to either cancel or drain your pipeline and then select Stop job.

5. Move to the BigQuery page within the Google Cloud Console.

6. In the Explorer panel, expand your project to view the associated datasets.

7. Next to the dataset you want to delete, click the three vertical dots (more view actions) and select Open.

8. In the Details panel, click Delete dataset and follow the instructions to confirm the deletion.

9. Next, go to the Cloud Storage browser page within the Google Cloud Console.

10. Click the checkbox next to the bucket that you want to delete. To delete the bucket, click Delete and follow the instructions to complete the deletion process.

Summary

In this chapter, we delved into the realm of data processing and transformation, focusing on Google Cloud Dataflow and Google Cloud Dataprep.

We began with an introduction to Google Cloud Dataflow, a powerful service for batch and stream data processing. We explored the various use cases where Dataflow shines, such as real-time analytics, ETL (Extract, Transform, Load) pipelines, and event-driven processing. We gained an understanding of how Dataflow enables scalable and efficient data processing.

Next, we dived into running data processing pipelines on Cloud Dataflow. We learned about the concept of pipelines, which allows us to define the series of steps needed to process and transform our data. We explored how Dataflow supported batch and stream processing and discovered its integration with other Google Cloud services, such as BigQuery and Cloud Pub/Sub.

Moving on, we explored Google Cloud Dataprep, a data preparation and transformation service. We learned about the various use cases where Dataprep can be applied, such as data cleansing, normalization, and data wrangling. We discovered the intuitive visual interface of Dataprep, which allows users to easily explore and transform their data without writing complex code.

Throughout the chapter, we gained insights into the importance of data processing and transformation in the data lifecycle. We learned how Google Cloud Dataflow and Google Cloud Dataprep provide powerful tools to process and prepare data at scale, enabling us to extract valuable insights and drive informed decision-making.

By harnessing the potential of Dataflow and Dataprep, we gain the expertise and proficiency to design and execute efficient data processing pipelines. We can seamlessly convert raw data into a usable format, guaranteeing data quality and consistency. This enables us to unleash the complete potential of our data and extract valuable insights that propel business growth and achievement.

CHAPTER 6

Data Analytics and Storage

This chapter delves into data analytics and storage, spotlighting Google Cloud Storage, Google Cloud SQL, and Google Cloud Pub/Sub. We commence with Google Cloud Storage, a scalable object storage solution. It finds utility in data archiving, hosting websites, and backups. We explore its features and compatibility with other Google services.

Google Cloud SQL follows a managed relational database service. It excels in storing relational data with automatic backups, scalability, and compatibility with engines like MySQL and PostgreSQL. Its use spans web apps, content management, and analytics.

Moving on, Google Cloud Pub/Sub enables real-time data streaming. We learn its core concepts and applications in event-driven systems, real-time analytics, and IoT. Integration with Cloud Functions and Dataflow is also examined.

We conclude by mastering data streams with Cloud Pub/Sub. We grasp topics, message publication, and subscriber setup. The focus on scalability and reliability empowers robust data streaming pipelines.

This chapter underscores the data's significance and the tools provided by Google Cloud Storage, Cloud SQL, and Cloud Pub/Sub to derive insights and make informed decisions. This knowledge equips us to design storage solutions, manage databases, and utilize real-time data effectively.

© Shitalkumar R. Sukhdeve and Sandika S. Sukhdeve 2023
S. R. Sukhdeve and S. S. Sukhdeve, *Google Cloud Platform for Data Science*,
https://doi.org/10.1007/978-1-4842-9688-2_6

Introduction to Google Cloud Storage and Its Use Cases for Data Storage

Google Cloud Storage is a highly scalable and reliable object storage service offered by Google Cloud Platform (GCP). It provides durable and highly available storage for unstructured data that can be accessed from anywhere in the world via the Internet. Here are some common use cases for Google Cloud Storage:

1. **Data backup and disaster recovery**: Google Cloud Storage can be used to store backups of critical data and provide disaster recovery solutions for organizations.

2. **Data archiving**: Google Cloud Storage provides a cost-effective solution for archiving large amounts of infrequently accessed data for long-term retention.

3. **Serving static content for websites and mobile applications**: Google Cloud Storage can serve static content like images, videos, and audio files for websites and mobile applications.

4. **Content distribution and media streaming**: Google Cloud Storage can be used to distribute and stream media files, enabling faster downloads and a better user experience.

5. **Data analytics and machine learning**: Google Cloud Storage can be integrated with other GCP services like BigQuery and TensorFlow to store, process, and analyze large volumes of data.

6. **Collaborative workflows**: Google Cloud Storage
   can be used for shared access to files, enabling
   teams to work together on projects and store and
   share data across different devices and locations.

Google Cloud Storage provides a highly scalable and reliable solution for storing and accessing unstructured data, making it a suitable choice for a wide range of use cases.

Key Features

Google Cloud Storage is a highly scalable, durable, and secure object storage service that provides a range of features to help users store, manage, and access their data. Here are some of the key features of Google Cloud Storage:

Scalability: Google Cloud Storage provides virtually unlimited storage capacity, allowing users to store and manage petabytes of data.

Durability: Google Cloud Storage provides high durability by automatically replicating data across multiple geographic locations, ensuring data is protected against hardware failures and other types of disasters.

Security: Google Cloud Storage provides multiple layers of security, including encryption at rest and in transit, fine-grained access controls, and identity and access management (IAM).

Performance: Google Cloud Storage provides high-performance data access, enabling fast upload and download speeds for objects of any size.

Integration with other Google Cloud Platform services: Google Cloud Storage integrates seamlessly with other GCP services, such as Compute Engine, App Engine, and BigQuery, to support a wide range of use cases.

Multi-Regional and Regional storage options: Google Cloud Storage offers Multi-Regional and Regional storage options to cater to different data storage and access needs.

Object Lifecycle Management: Google Cloud Storage supports Object Lifecycle Management, allowing users to define rules to automate data retention, deletion, or archival policies.

Cost-effectiveness: Google Cloud Storage offers a flexible pricing model that allows users to pay only for what they use, making it a cost-effective option for storing and accessing data.

Google Cloud Storage offers a wide range of features that make it a highly scalable, durable, and secure object storage service suitable for a variety of use cases.

Storage Options

Google Cloud Storage provides multiple storage options that cater to different data storage and access needs. Here are some of the storage classes offered by Google Cloud Storage:

1. **Standard**: This storage class is suitable for frequently accessed data, such as active application data and user data, with low latency and high durability.

2. **Nearline**: This storage class is ideal for data accessed less frequently but that must be available within seconds, such as monthly financial reports or large media files.

3. **Coldline**: This storage class is designed for rarely accessed data but that needs to be available within minutes, such as disaster recovery data or long-term archives.

4. **Archive**: This storage class is intended for rarely accessed data and has a long retention period, such as regulatory archives or compliance records.

Storage Locations

In addition to the preceding storage classes, Google Cloud Storage also provides two types of storage locations:

1. **Regional storage**: This option stores data in a specific region, allowing for lower-latency access to data within the region. Regional storage is recommended for applications with high-performance requirements.

2. **Multi-Regional storage**: This option stores data in multiple geographic locations, providing high availability and global access to data. Multi-Regional storage is recommended for applications that require global access to data.

The multiple storage options provided by Google Cloud Storage allow users to choose the right storage class and location that meet their data storage and access needs and optimize their storage costs.

Creating a Data Lake for Analytics with Google Cloud Storage

It is a straightforward process that involves the following steps:

1. **Create a Google Cloud Storage bucket**: A bucket is a container for storing objects, such as files or data. You can create a bucket in the GCP Console or use the gsutil command-line tool. Open the Cloud Console navigation menu and then select Cloud Storage.

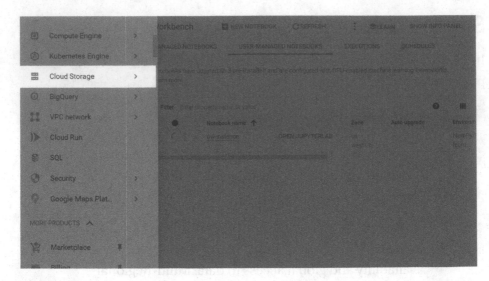

Figure 6-1. *Cloud Storage*

You will be able to see the following screen.

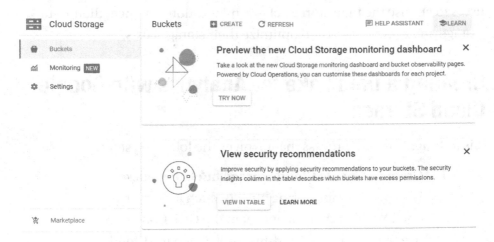

Figure 6-2. *Bucket*

Next, click CREATE.

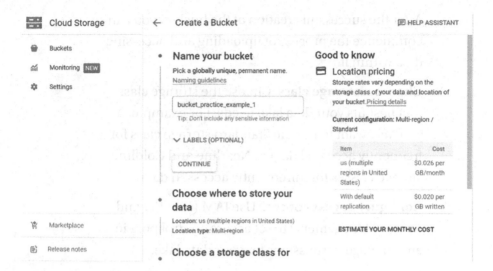

Figure 6-3. *New bucket creation*

Provide a distinct name for the bucket, select the
preferred region, and opt for the desired storage
class, such as standard or infrequent access. Adjust
any additional configurations, such as versioning
and lifecycle rules. Finally, initiate the creation of
the bucket by clicking the Create button.

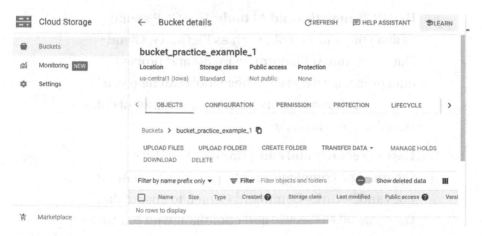

Figure 6-4. *Bucket details*

After the successful creation of the bucket, you can commence the process of uploading and accessing data within it.

2. **Choose a storage class**: Choose the storage class that best fits your data lake needs. For example, you may want to use the Standard storage class for frequently accessed data or Nearline and Coldline storage classes for infrequently accessed data.

3. **Configure access control**: Use IAM (Identity and Access Management) to set access control policies and manage permissions for your data lake.

4. **Ingest data into the data lake**: You can upload data to your bucket using the GCP Console, the gsutil command-line tool, or other supported data ingestion tools.

5. **Organize data in the data lake**: Use a directory structure or object naming conventions to organize data within your bucket. This will make it easier to search and access data when performing analytics.

6. **Use GCP analytics and AI tools**: Use GCP's suite of analytics and AI tools, such as BigQuery, Cloud Dataflow, and AI Platform, to ingest and process data from your data lake. These tools can help you analyze and gain insights from large volumes of data stored in your data lake.

7. **Ensure security and compliance**: Use GCP's security and compliance tools, such as Cloud Audit Logs and Security Command Center, to ensure that your data lake is secure and compliant with industry regulations.

Creating a data lake for analytics with Google Cloud Storage involves setting up a bucket, configuring access control, ingesting and organizing data, and using GCP's analytics and AI tools to gain insights from the data. By adhering to these instructions, you can establish a robust data lake that empowers you to make informed decisions based on data and gain a competitive edge within your industry.

Introduction to Google Cloud SQL and Its Use Cases for Relational Databases

Google Cloud SQL is a fully managed relational database service provided by Google Cloud Platform (GCP) that enables users to create, configure, and use databases in the cloud. With Google Cloud SQL, users can leverage the power and flexibility of popular relational database management systems, such as MySQL, PostgreSQL, and SQL Server, without the need to manage the underlying infrastructure, patching, or backups.

Google Cloud SQL provides several features that make it a powerful and flexible database solution, including the following:

Automatic backups and point-in-time recovery: Google Cloud SQL automatically backs up databases and enables point-in-time recovery, allowing users to easily restore databases to a specific point in time.

High availability and durability: Google Cloud SQL provides a highly available and durable database solution with automated failover and built-in replication across multiple zones and regions.

Scalability: Google Cloud SQL allows users to scale their database resources up or down as needed, providing flexibility and cost-effectiveness.

Compatibility: Google Cloud SQL is compatible with popular open source and commercial relational database management systems, making migrating existing databases to the cloud easy.

Security: Google Cloud SQL provides multiple layers of security, including encryption at rest and in transit, IAM-based access control, and network security (Secureicon, 2022).

Integration with other GCP services: Google Cloud SQL integrates seamlessly with other GCP services, such as Compute Engine, App Engine, and Kubernetes Engine, enabling users to build and deploy powerful applications in the cloud.

Overall, Google Cloud SQL provides a fully managed, highly available, and scalable database solution that enables users to easily deploy, manage, and use popular relational database management systems in the cloud.

Create a MySQL Instance by Using Cloud SQL

1. To initiate the process, access the Google Cloud Console at `https://console.cloud.google.com/` and log in. Then, choose the desired project in which you intend to create the MySQL instance.

2. Enable APIs:

    Compute Engine API

    sqladmin API (prod)

3. Once you're in the project, click the navigation menu in the top-left corner and select SQL.

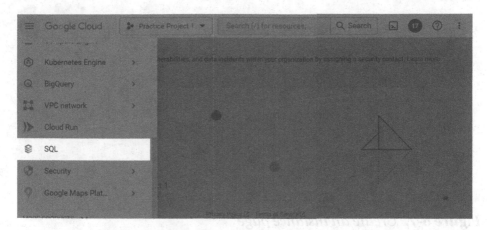

Figure 6-5. *SQL menu item*

4. Click the CREATE INSTANCE button on the Cloud
    SQL page.

Figure 6-6. *CRATE INSTANCE*

5. On the Create an instance page, select the MySQL
    version you want to use.

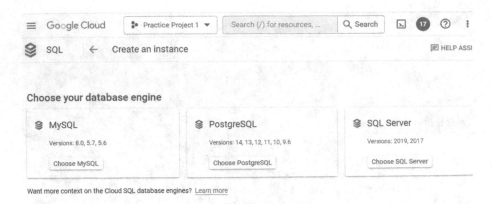

Figure 6-7. *Create an instance page*

6. Choose the instance type and configure the instance settings as per your requirements. You can set the instance ID, region, and zone where you want to create the instance.

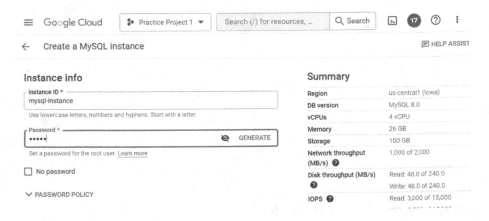

Figure 6-8. *Configure the instance settings*

7. Next, configure the machine type and storage type for the instance. You can also set the storage capacity.

8. Under the Connectivity section, you can choose how to connect to your instance. You can choose a public or private IP address and configure authorized networks for access.

9. Under the Additional Configuration section, you can configure advanced options such as database flags, backup settings, and maintenance settings.

10. Once you've configured all the settings, click the CREATE INSTANCE button at the bottom of the page to create the MySQL instance.

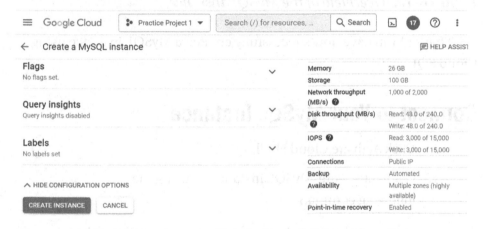

Figure 6-9. *Create a MySQL instance*

11. The instance creation process may take a few minutes, depending on the configuration and resources allocated. Once the instance is created, connect to it using any MySQL client and start using it.

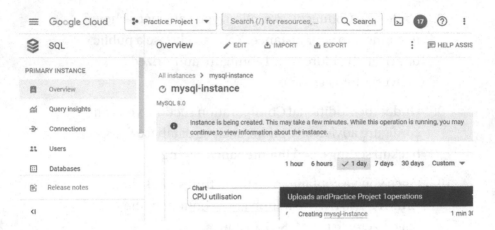

Figure 6-10. *Overview of the MySQL instance*

That's it! You have now successfully created a MySQL instance using Cloud SQL.

Connect to Your MySQL Instance

1. Click Activate Cloud Shell.

2. Connect to your MySQL instance by using the following command:

   gcloud sql connect mysql-instance --user=root

Figure 6-11. *Connect to the MySQL instance*

3. Authorize.

Figure 6-12. *Authorize Cloud Shell*

4. Enter the password created while creating the
   SQL instance, and you will get a MySQL prompt
   (mysql>).

Create a Database and Upload Data in SQL

1. This code will create a new database named "college." You can replace "college" with the name you want to give to your database:

    CREATE DATABASE college;

    Remember to use appropriate naming conventions for your database to make it easily identifiable and to avoid naming conflicts with other databases.

2. The following code will first select the "college" database using the USE statement and then create a new table named "students" with three columns: "id", "name", and "age". The "id" column is an auto-incrementing integer that serves as the primary key for the table:

    USE college;

    ```
    CREATE TABLE students (
        id INT NOT NULL AUTO_INCREMENT,
        name VARCHAR(50) NOT NULL,
        age INT NOT NULL,
        PRIMARY KEY (id)
            );
    ```

    You can modify the column names, data types, and constraints to fit your specific needs. The CREATE TABLE statement provides a lot of flexibility in defining the structure of your database tables.

3. The following code will insert five rows of data into
    the "students" table with values for the "name" and
    "age" columns. The "id" column is not specified,
    since it is an auto-incrementing primary key that
    will be generated automatically by MySQL:

```
INSERT INTO students (name, age) VALUES
('John Smith', 20),
('Jane Doe', 21),
('Bob Johnson', 19),
('Samantha Lee', 22),
    ('Chris Brown', 18);
```

You can modify the sample data to fit your specific
needs and insert as many rows as you need.
Remember to provide values for all the columns that
are not defined as AUTO_INCREMENT or DEFAULT
NULL, or you will get an error.

Note that the INSERT INTO statement can also be
used to insert data into multiple tables at once or to
insert data from another table using a subquery.

Figure 6-13. *INSERT INTO*

4. View the data from the table:

    *select * from students;*

Figure 6-14. *select*

5. Clean up.

    Use the following code to delete the database:

    DROP DATABASE college;

6. Disconnect from your MySQL instance using the
    exit command and close the shell.

7. Delete the instance. If the delete option is not
    enabled, go to the Data Protection section and
    disable deletion protection.

Introduction to Google Cloud Pub/Sub and Its Use Cases for Real-Time Data Streaming

Google Cloud Pub/Sub is a messaging service provided by Google
Cloud Platform that enables asynchronous communication between
applications. It allows developers to build highly scalable and reliable
messaging systems to exchange messages between different components
of their applications in a decoupled and fault-tolerant manner.

The basic idea behind Pub/Sub is that publishers send messages to topics and subscribers receive messages from these topics. Topics are virtual channels to which messages can be sent, and subscribers can be attached to one or more topics to receive messages from them. Messages can be in any format, including JSON, XML, or binary data, and can be up to 10 MB in size.

Google Cloud Pub/Sub provides a range of functionalities that simplify the utilization and scalability of messaging systems for developers. For example, it provides a REST API and client libraries in several programming languages, including Java, Python, and Go. It also provides durable message storage, allowing messages to be stored and delivered to subscribers even if they are offline or unavailable.

Pub/Sub also integrates with other Google Cloud services, such as Google Cloud Functions, Google Cloud Dataflow, and Google Cloud Storage, to provide a seamless experience for developers building cloud-native applications. Additionally, Pub/Sub is highly scalable and can handle millions of messages per second, making it an ideal choice for high-throughput messaging scenarios.

Google Cloud Pub/Sub is widely used for real-time data streaming use cases, such as

1. **IoT data ingestion**: Pub/Sub can be used to ingest and process large volumes of real-time data generated by IoT devices. Devices can publish data to topics, and subscribers can consume the data to perform real-time analysis or trigger actions based on the data.

2. **Clickstream processing**: Pub/Sub can be used to process clickstream data generated by websites or mobile apps in real time. The data can be streamed to Pub/Sub topics, and subscribers can consume the data to analyze user behavior and make real-time decisions.

179

3. **Financial trading**: Pub/Sub can be used to stream real-time stock quotes, financial news, and other market data to subscribers, allowing them to react quickly to market changes and execute trades in real time.

4. **Gaming**: Pub/Sub can be used to stream real-time gaming events, such as player movements, game state changes, and in-game purchases. This allows developers to build real-time multiplayer games and deliver personalized gaming experiences to users.

5. **Real-time analytics**: Pub/Sub can be used to stream real-time data to analytics platforms such as Google Cloud Dataflow or Apache Beam, allowing real-time analysis and decision-making.

In summary, Google Cloud Pub/Sub provides a reliable and scalable messaging service that can be used for real-time data streaming use cases in various industries, such as IoT, finance, gaming, and analytics.

Setting Up and Consuming Data Streams with Cloud Pub/Sub

Hands-on Pub/Sub step-by-step:

1. **The project setup**: To make sure you have the necessary rights, either create a new project or choose an existing project in which you already have them.

2. **Create a topic**: Go to the Pub/Sub browser to create a topic.

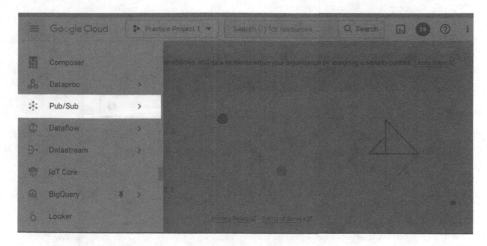

Figure 6-15. *Pub/Sub*

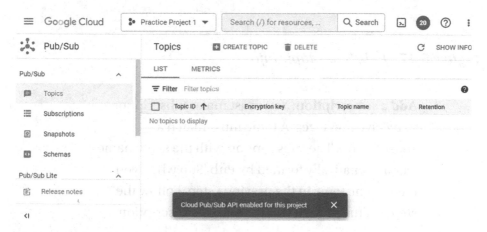

Figure 6-16. *Pub/Sub—CREATE TOPIC*

Click CREATE TOPIC. Fill in the required information and click CREATE TOPIC.

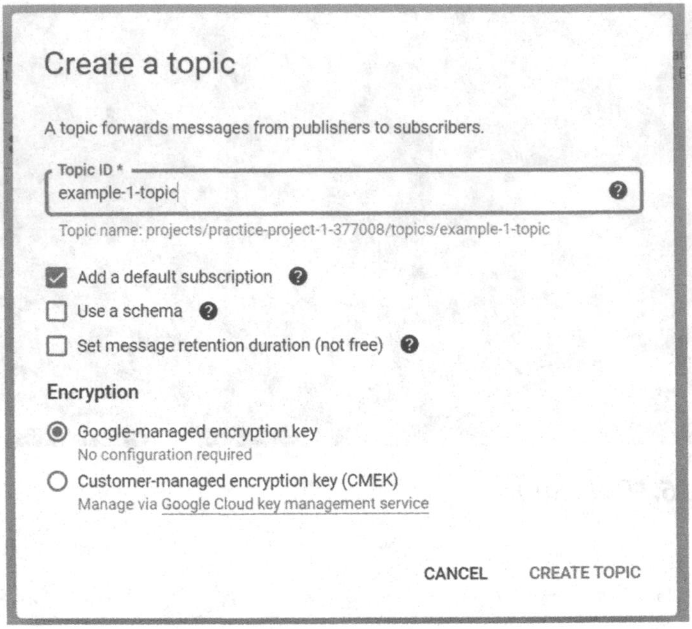

Figure 6-17. *Pub/Sub—topic info*

3. **Add a subscription**: You must make subscriptions
   to receive messages. A topic must match a
   subscription. The subscription with the same name
   was automatically formed by Pub/Sub when you
   created the topic in the previous step. Follow the
   steps on the page to create another subscription.

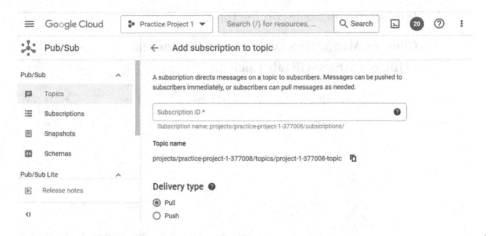

Figure 6-18. *Subscription creation*

Scroll down the page and click Create after filling in
the necessary information.

4. Return to the topic page and click the topic.
Scrolling down, you should be able to see the
subscription created.

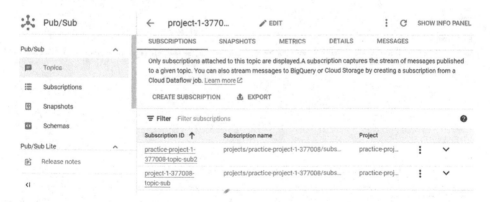

Figure 6-19. *Pub/Sub—subscription creation*

5. Publish two messages on the topic.

Click the MESSAGES tab, and you can find the
PUBLISH MESSAGE tab. Click it.

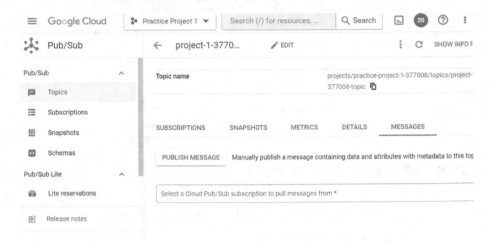

Figure 6-20. *Publish messages*

Publish message

Topic name
projects/practice-project-1-377008/topics/project-1-377008-topic

Publish count
You can publish the given message once or multiple times in an interval. This can be useful for getting messages in new subscriptions and testing. For a more robust way to publish messages multiple times, consider using Cloud Scheduler.

Number of messages *
1

Enter an amount between 1 and 100.

Message interval (seconds)
1

How long to wait before publishing the next message

Message body
The message that you want to publish to this topic. Either message or attribute will be required to publish.

Message *
Hi, my first Pub/Sub message.

PUBLISH   CANCEL

Figure 6-21. *Publish a message*

Enter a message in the message body.

6. Go to the subscription page and click the
    subscription. Go to MESSAGES. If you don't find
    any message, click PULL till you get the message.
    You should be able to see the message in the
    message body.

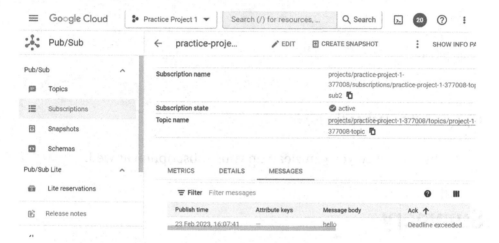

Figure 6-22. Message pull

Pub/Sub does not guarantee the messages'
chronological sequence. You might have only
noticed one statement. If so, keep clicking PULL
until the other notification appears.

7. **Clean up**: The topic you created can be deleted
    if you want to tidy up the Pub/Sub resources you
    made in this tutorial. Visit the Topics tab once
    more. Click DELETE after selecting the box next to
    the topic.

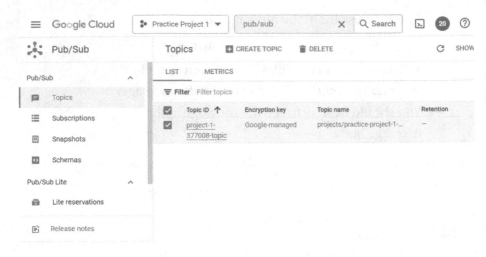

Figure 6-23. *Delete the topic*

In the same way, you can clean up your subscription as well.

Summary

This chapter explored the world of data analytics and storage, focusing on Google Cloud Storage, Google Cloud SQL, and Google Cloud Pub/Sub.

We began by introducing Google Cloud Storage, a highly scalable and durable object storage service. We discussed the various use cases where Cloud Storage is beneficial, such as storing and archiving data, hosting static websites, and serving as a backup and recovery solution. We explored the key features of Cloud Storage, including its storage classes, access controls, and integration with other Google Cloud services.

Next, we delved into Google Cloud SQL, a fully managed relational database service. We discussed the advantages of using Cloud SQL for storing and managing relational data, including its automatic backups, scalability, and compatibility with popular database engines such as MySQL and PostgreSQL. We explored the use cases where Cloud SQL is suitable, such as web applications, content management systems, and data analytics.

Moving on, we turned our attention to Google Cloud Pub/Sub, a messaging service for real-time data streaming. We learned about the core concepts of Pub/Sub, including topics, publishers, and subscribers. We discovered the use cases where Pub/Sub excels, such as event-driven architectures, real-time analytics, and IoT data processing. We also explored the integration of Pub/Sub with other Google Cloud services, such as Cloud Functions and Dataflow.

Finally, we learned how to set up and consume data streams with Cloud Pub/Sub. We explored the steps involved in creating topics, publishing messages, and creating subscribers to receive and process the streamed data. We gained insights into the scalability and reliability of Pub/Sub, making it an ideal choice for building robust and scalable data streaming pipelines.

Throughout the chapter, we gained a deeper understanding of the importance of data storage and analytics in the modern data landscape. We learned how Google Cloud Storage, Cloud SQL, and Cloud Pub/Sub provide powerful tools and services to store, manage, and analyze data at scale. By leveraging these services, organizations can unlock the value of their data, derive meaningful insights, and make data-driven decisions.

With the knowledge and skills acquired from this chapter, we are equipped to design and implement robust data storage solutions, leverage relational databases for efficient data management, and harness the power of real-time data streaming for real-time analytics and event-driven architectures.

## CHAPTER 7

Advanced Topics

This chapter delves into advanced aspects of securing and managing Google Cloud Platform (GCP) resources, version control using Google Cloud Source Repositories, and powerful data integration tools: Dataplex and Cloud Data Fusion.

We start by emphasizing secure resource management with Identity and Access Management (IAM), covering roles, permissions, and best practices for strong access controls.

Google Cloud Source Repositories is explored as a managed version control system for scalable code management and collaboration. Dataplex is introduced as a data fabric platform, aiding data discovery, lineage, and governance, while Cloud Data Fusion simplifies data pipeline creation and management.

Throughout, we'll gain insights to enhance GCP security, streamline code collaboration, and optimize data integration for various processing needs.

Securing and Managing GCP Resources with IAM

IAM (Identity and Access Management) is a service offered by Google Cloud Platform (GCP) that facilitates the administration of access to your GCP resources. IAM empowers you to regulate resource accessibility

© Shitalkumar R. Sukhdeve and Sandika S. Sukhdeve 2023
S. R. Sukhdeve and S. S. Sukhdeve, *Google Cloud Platform for Data Science*,
https://doi.org/10.1007/978-1-4842-9688-2_7

and define the actions that individuals can execute on those resources. It provides a centralized place to manage access control, making enforcing security policies easier and reducing the risk of unauthorized access.

Here are some key concepts related to IAM:

1. **Resource**: A GCP entity such as a project, bucket, or instance that IAM policies can be applied.

2. **Role**: Roles are a collection of permissions that specify the actions permitted on a particular resource. These roles can be assigned to users, groups, or service accounts, determining the scope of their operations.

3. **Member**: An identity that can be assigned a role. Members can be users, groups, or service accounts.

4. **Policy**: A set of role bindings that define who has what permissions on a resource.

Here are some key features of IAM:

1. **Role-Based Access Control (RBAC)**: IAM uses RBAC to control access to resources. RBAC allows you to define roles and assign them to users, groups, or service accounts.

2. **Custom roles**: IAM provides predefined roles with predefined permissions, but you can create custom roles that fit your specific needs. This allows you to grant more granular permissions to users or services.

3. **Service accounts**: Service accounts are a particular type of account that can be used by applications or services to authenticate with GCP APIs. They can be given permissions just like human users but are not tied to a specific person.

4.   **Audit logging**: IAM provides audit logging that allows you to track who has access to your resources and their actions.

5.   **Fine-grained access control**: IAM provides fine-grained access control over resources. You can control access to specific resources within a project and even within a particular folder or organization.

6.   **Conditional access**: IAM provides conditional access, allowing you to set conditions for users or services accessing your resources. For example, you can set conditions based on the IP address or device.

In summary, IAM is a powerful service that allows you to manage access to your GCP resources. Using IAM, you can enforce security policies, reduce the risk of unauthorized access, and ensure your resources are well-managed.

Using the Resource Manager API, Grant and Remove IAM Roles

1.   To make sure you have the necessary permissions, either create a new project or choose a current project in which you already have them.

2.   Enable APIs:

     Identity and Access Management (IAM) API and Cloud Resource Manager API

3.   Grant a principal the Logs Viewer role on the project.

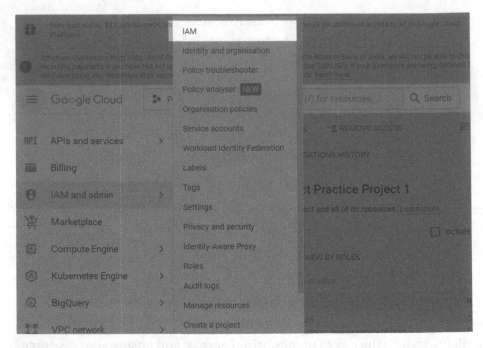

Figure 7-1. *IAM and admin*

4.   Click GRANT ACCESS.

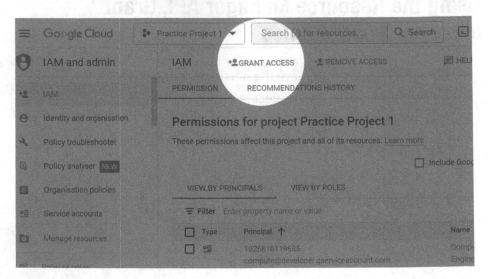

Figure 7-2. *GRANT ACCESS*

5. Enter the email address of a principal.

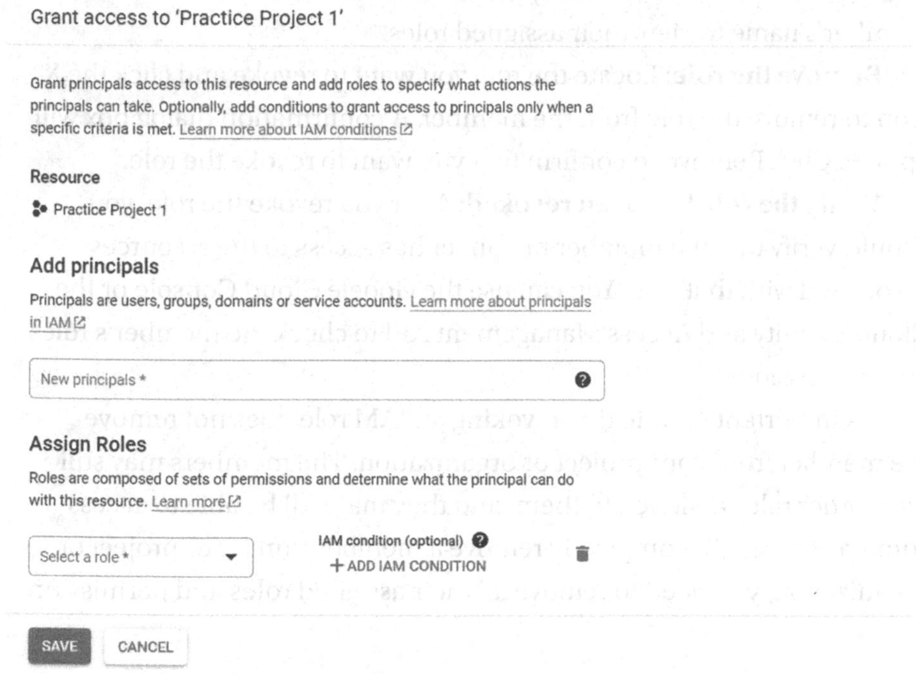

Grant access to 'Practice Project 1'

Grant principals access to this resource and add roles to specify what actions the principals can take. Optionally, add conditions to grant access to principals only when a specific criteria is met. Learn more about IAM conditions ☑

Resource

• Practice Project 1

Add principals

Principals are users, groups, domains or service accounts. Learn more about principals in IAM ☑

New principals *   ❷

Assign Roles

Roles are composed of sets of permissions and determine what the principal can do with this resource. Learn more ☑

Select a role *   ▼   IAM condition (optional) ❷   🗑
                              + ADD IAM CONDITION

SAVE   CANCEL

Figure 7-3. *GRANT ACCESS—info*

6. From the Select a role menu, select the product or service you want to give access to.

7. If you want to revoke access, you can do this by following the following steps:

Identify the member and the role to be revoked: Before you can revoke an IAM role, you need to know which member has been assigned the role and which role you want to revoke.

Navigate to the IAM page: Open the IAM page in the Google Cloud Console by clicking the IAM tab in the left navigation menu.

Find the member: Use the search bar at the top of the page to find the member who has been assigned the role you want to revoke. Click the member's name to view their assigned roles.

Remove the role: Locate the role you want to revoke and click the X icon to remove the role from the member. A confirmation dialog box will appear. Click Remove to confirm that you want to revoke the role.

Verify the role has been revoked: After you revoke the role, you should verify that the member no longer has access to the resources associated with that role. You can use the Google Cloud Console or the Cloud Identity and Access Management API to check the member's roles and permissions.

It's important to note that revoking an IAM role does not remove the member from your project or organization. The members may still have other roles assigned to them, and they may still be able to access some resources. To completely remove a member from your project or organization, you need to remove all their assigned roles and permissions.

Using Google Cloud Source Repositories for Version Control

Google Cloud Source Repositories is a source code management service provided by Google Cloud Platform (GCP). It allows you to store and manage your source code in a secure and scalable way and provides advanced features such as code search, code reviews, and continuous integration and deployment (CI/CD).

Here are some key features of Google Cloud Source Repositories:

> **Scalability**: Google Cloud Source Repositories can handle repositories of any size and can scale to meet the needs of large enterprises.

Code search: Google Cloud Source Repositories provides advanced code search capabilities that allow you to search across all your repositories for specific code snippets, files, or even functions.

Code reviews: Google Cloud Source Repositories makes it easy to review code changes and collaborate with other developers. You can create pull requests, assign reviewers, and track changes in real time.

Integrated CI/CD: Google Cloud Source Repositories integrates with other GCP services, such as Cloud Build and Cloud Functions, allowing you to build and deploy your code automatically.

Secure: Google Cloud Source Repositories provides multiple layers of security to protect your code, including encryption at rest and in transit, access controls based on IAM roles, and audit logs.

Flexible: Google Cloud Source Repositories supports a wide range of source code management tools and protocols, including Git, Mercurial, and Subversion.

Google Cloud Source Repositories is fully integrated with other GCP services, allowing you to use it seamlessly with other GCP tools and services. It's a powerful solution for managing source code, collaborating with other developers, and automating software development workflows.

Dataplex

Dataplex is a new data platform launched by Google Cloud in 2021. It is designed to simplify the complexities of modern data management by providing a unified, serverless, and intelligent data platform. Dataplex allows users to manage their data across multiple clouds and on-premises data stores through a single interface.

The platform offers many key features that make it a valuable tool for data management:

1. **Unified data fabric**: Dataplex provides a unified data fabric that enables users to manage their data across multiple data stores, including Google Cloud Storage, BigQuery, Cloud Spanner, and Cloud SQL. This allows users to store and manage their data in the most appropriate data store for their use case while maintaining a consistent view of their data.

2. **Serverless architecture**: Dataplex is built on a serverless architecture, which means that users don't have to worry about managing infrastructure. This makes setting up and scaling data management systems easy while reducing costs and increasing agility.

3. **Intelligent data management**: By employing machine learning algorithms, Dataplex automates various data management tasks including data classification, data lineage, and data discovery. This automation enhances the effective governance and management of data, mitigating the chances of errors and improving overall efficiency.

4. **Open platform**: Dataplex is an open platform
    that allows users to use their preferred tools and
    technologies. This makes it easy to integrate with
    existing systems and workflows and to extend the
    platform with custom applications and services.

Dataplex is a powerful data management platform that simplifies
the complexities of modern data management. With its diverse range of
features, Dataplex proves to be an invaluable tool for data scientists, data
engineers, and other professionals dealing with extensive and intricate
datasets.

This tutorial will guide you through the fundamental steps of
configuring Dataplex within the Google Cloud Console.

Step 1: Create a Google Cloud project.

1. First, you must create a new project in the Google
    Cloud Console. However, if you already have a
    project, please proceed to the next step.

2. Go to the Google Cloud Console.

3. Click the project drop-down at the top of the page
    and select NEW PROJECT.

4. Enter a name for your project and click CREATE.

Step 2: Enable the Dataplex API.

Next, we need to enable the Dataplex API for our project.

1. In the Google Cloud Console, navigate to APIs and
    services.

2. Click the + Enabled APIs and services button.

Figure 7-4. *+ Enabled APIs and services*

3. Search for "Dataplex" in the search bar and select Dataplex API from the results.

Figure 7-5. *Select the Dataplex API*

4. Click the ENABLE button to enable the API.

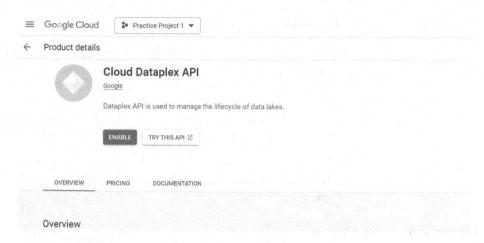

Figure 7-6. Enable Dataplex

Step 3: Create a lake.

1.   Go to Dataplex in the cloud console and click
     Explore.

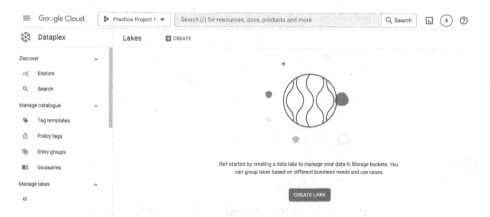

Figure 7-7. Dataplex—create a lake

2.   After clicking CREATE LAKE, you can see the
     following image. Fill in the required details.
     Click CREATE.

Figure 7-8. *Create a lake*

3. After creating a data lake, you can see the following screen.

Figure 7-9. *Data lake created*

4. Once you have created your lake, it is possible to add zones. These zones serve as a means of categorizing unstructured and structured data logically. Click the name of lake, and you can see the following screen.

Figure 7-10. *Manage the lake*

5. Click +ADD ZONE, and you can see the
    following screen.

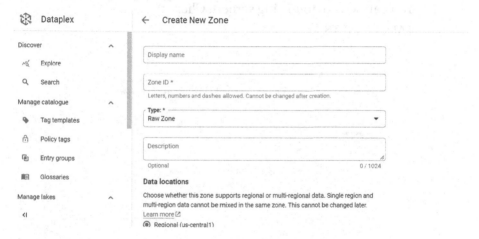

Figure 7-11. *Create a zone*

6. Once the zone is added, we can add an asset to the
    lake by clicking the zone.

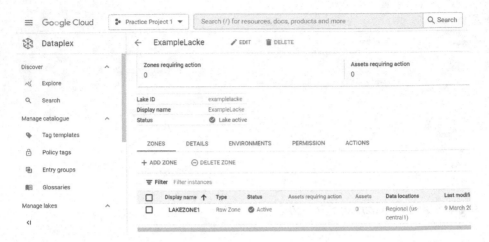

Figure 7-12. *Lake zone created*

7.   You can see the following screen. Click
     +ADD ASSETS.

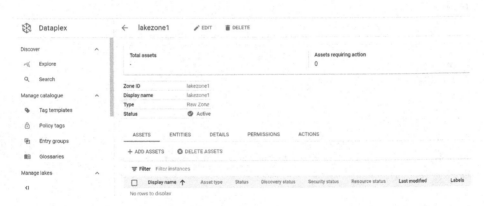

Figure 7-13. *ADD ASSETS—screen 1*

8.   From the following image, you can click ADD
     AN ASSET.

Figure 7-14. *ADD ASSETS—screen 2*

9. Fill in the information in the following form.

Figure 7-15. *Add asset information*

10. Here we are adding a BigQuery dataset. Give
    a display name that will be the same as the ID
    automatically. Scroll down and browse for the
    dataset. A list of datasets will appear. Select the
    dataset. Click Done and then Continue.

11. Discovery settings will appear. Click Continue.

12. Review your asset and click Submit.

13. You can see the following image if everything is successful.

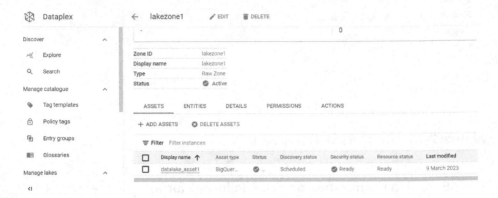

Figure 7-16. *Asset status*

14. For not getting any cost, if you don't need the resources created, they should be deleted. First, delete the assets. Then you can delete the zones and then the lake.

Cloud Data Fusion

Cloud Data Fusion is a fully managed data integration service that allows users to efficiently build and manage ETL/ELT (Extract, Transform, Load/ Extract, Load, Transform) pipelines on Google Cloud Platform. The service is designed to make it easy for data analysts and data engineers to create, deploy, and manage data pipelines at scale without requiring deep technical expertise.

Cloud Data Fusion comes with a web-based UI that provides a graphical interface for building and managing data pipelines. The UI includes a drag-and-drop interface for connecting sources and destinations and a large library of pre-built connectors to common data sources and sinks. Cloud Data Fusion also supports complex data transformations, such as join, merge, and filter operations.

Under the hood, Cloud Data Fusion uses Apache Spark and Apache Beam, two open source data processing frameworks, to execute data transformations and move data between systems. This allows Cloud Data Fusion to support a wide range of data sources, including relational databases, NoSQL databases, and file systems, and to scale to handle large data volumes.

With Cloud Data Fusion, users can easily create data pipelines to move data between various systems, such as from on-premises data centers to Google Cloud or between different services on Google Cloud. Cloud Data Fusion integrates with other Google Cloud services such as BigQuery, Cloud Storage, and Cloud Pub/Sub, making moving data between these services easy.

Overall, Cloud Data Fusion can simplify the process of building and managing data pipelines, allowing users to focus on extracting value from their data instead of managing the underlying infrastructure.

Enable or Disable Cloud Data Fusion

Cloud Data Fusion can be used with an existing project, or a new one can be created. In the case of an existing project, the Cloud Data Fusion API needs to be enabled.

To enable the Cloud Data Fusion API and create a new project, follow these steps:

1. Select or create a project in the Google Cloud Console.

2. Ensure that billing is enabled for your project.

3. Navigate to the API Overview page for Cloud Data Fusion.

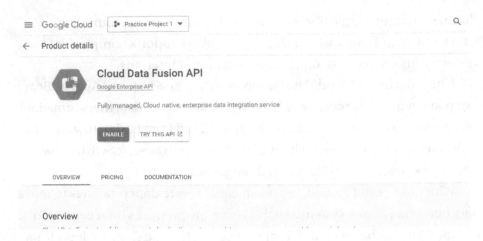

Figure 7-17. *Cloud Data Fusion*

4. Click the ENABLE button to activate the Cloud Data Fusion API.

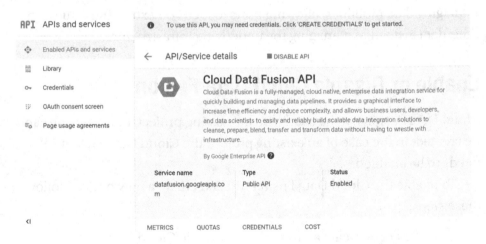

Figure 7-18. *API enabled*

You can start using Cloud Data Fusion with your project by completing the following steps.

Create a Data Pipeline

1. Go to Create Data Fusion instance.

2. Enter an instance name.

3. If you are using Cloud Data Fusion version 6.2.3 or later, select a Dataproc service account to use for running your pipeline in Dataproc from the Authorization field. The default option is the Compute Engine account.

   Once you have provided all the required details, simply click the Create button. Please keep in mind that the instance creation process may take up to 30 minutes to finish. While Cloud Data Fusion is creating your instance, you will see a progress wheel next to the instance name on the Instances page. Once the process is complete, the wheel will transform into a green check mark, indicating that the instance is ready for use.

Figure 7-19. Data Fusion instance

4. Click View Instance.

Figure 7-20.   Cloud Data Fusion UI

5. Through the Cloud Data Fusion Hub, you can access
    sample pipelines, plugins, and solutions that are
    reusable and shareable among users of Cloud Data
    Fusion. Click **Hub**.

6. Click Pipelines in the bottom left.

Figure 7-21.   Hub Explorer

7. Click the Cloud Data Fusion Quickstart pipeline.
    Then click Finish and customize.

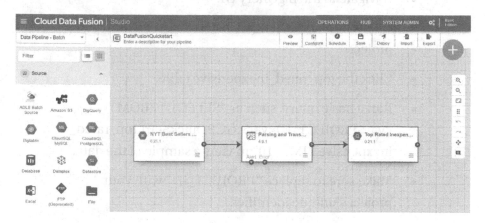

Figure 7-22. *Data Fusion Quickstart pipeline*

8. Click Deploy.

    Once the pipeline has been deployed, you can
    access its details view, where you can perform the
    following actions:

    - View the pipeline's structure and configuration.

    - Run the pipeline manually or configure a schedule
      or trigger for automated runs.

    - View a summary of the pipeline's past executions,
      including execution times, logs, and metrics.

9. Click **Run** to execute your pipeline.

10. Upon completion of a few minutes, the pipeline
    concludes and transitions its status to "Succeeded."
    Additionally, the number of records processed by
    each node is shown.

11. To view a sample of the results, follow these steps:

- Navigate to the BigQuery UI.

- Open the DataFusionQuickstart dataset within your project.

- Click the top_rated_inexpensive table.

- Run a basic query, such as "SELECT * FROM <var>PROJECT_ID<var>.GCPQuickStart.top_rated_inexpensive LIMIT 10" to view a sample of the data.

- Make sure to replace "PROJECT_ID" with your project's unique identifier.

12. To prevent your Google Cloud account from incurring charges related to the resources used on this page, please take the following actions:

- Remove the BigQuery dataset that your pipeline wrote to during this quickstart.

- Delete the Cloud Data Fusion instance.

- Please note that deleting your instance does not remove any data associated with your project.

- (Optional) If desired, you may also delete the entire project.

Summary

This chapter explored advanced topics related to securing and managing Google Cloud Platform (GCP) resources, version control with Google Cloud Source Repositories, and two powerful data integration and management tools: Dataplex and Cloud Data Fusion.

We began by discussing the importance of securing and managing GCP resources with Identity and Access Management (IAM). We learned about IAM roles, permissions, and policies and how they allow administrators to control access to GCP resources. By implementing strong IAM practices, organizations can ensure the security and integrity of their GCP resources.

Next, we delved into Google Cloud Source Repositories, a version control system that provides a scalable and fully managed repository for storing and managing source code.

Moving on, we explored Dataplex, a data fabric platform that enables organizations to discover, understand, and manage their data assets at scale. We discussed the key features of Dataplex, including data discovery, data lineage, and data governance. We learned how Dataplex integrates with other Google Cloud services, such as BigQuery and Cloud Storage, to provide a unified and efficient data management solution.

Finally, we learned about Cloud Data Fusion, a fully managed data integration service that simplifies the process of building and managing data pipelines. We discussed the capabilities of Cloud Data Fusion, including its visual interface for designing pipelines, support for a wide range of data sources and sinks, and built-in data transformation and cleansing capabilities. We also explored the integration of Cloud Data Fusion with other Google Cloud services, such as BigQuery and Dataproc, to create end-to-end data processing workflows.

Throughout the chapter, we gained insights into advanced topics that enhance our ability to secure and manage GCP resources effectively, collaborate on code development using version control systems, and leverage powerful data integration and management tools. By understanding and utilizing IAM, Cloud Source Repositories, Dataplex, and Cloud Data Fusion, we are equipped to tackle complex challenges related to resource management, version control, and data integration in the Google Cloud environment.

With the knowledge and skills acquired from this chapter, we can confidently navigate the advanced aspects of Google Cloud Platform, implement robust security measures, streamline code collaboration, and efficiently manage and integrate data for various data processing and analytics requirements.

Bibliography

Ahmed, F. (2023, January 25). *What is Tensorflow*. Retrieved from www.makermodules.com/what-is-tensorflow/

Blog, Neos Vietnam (n.d.). *Tutorial: TensorFlow Lite*. Retrieved from https://blog.neoscorp.vn/tutorial-tensorflow-lite/

Google Cloud (2023, February 7). *Introduction to Vertex AI*. Retrieved from https://cloud.google.com/vertex-ai/docs/start/introduction-unified-platform

Google Cloud (2023, June 6). *What is IaaS?* Retrieved from https://cloud.google.com/learn/what-is-iaas

Google Cloud (n.d.). *The Home Depot: Helping doers get more done through a data-driven approach*. Retrieved from https://cloud.google.com/customers/the-home-depot

Google Cloud (n.d.). *Spotify: The future of audio. Putting data to work, one listener at a time*. Retrieved from https://cloud.google.com/customers/spotify

IDC. (2021, June). *Worldwide Foundational Cloud Services Forecast, 2021–2025: The Base of the Digital-First Economy*. Retrieved from www.idc.com/getdoc.jsp?containerId=US46058920

Reis, N. (2016, May). *Deep Learning*. Retrieved from https://no.overleaf.com/articles/deep-learning/xhgfttpzrfkz

Secureicon. (2022). *Tag: cloud security incidents*. Retrieved from www.securicon.com/tag/cloud-security-incidents/

Sukhdeve, S. R. (2020). *Step Up for Leadership in Enterprise Data Science and Artificial Intelligence with Big Data: Illustrations with R and Python*. KDP.

© Shitalkumar R. Sukhdeve and Sandika S. Sukhdeve 2023
S. R. Sukhdeve and S. S. Sukhdeve, *Google Cloud Platform for Data Science*,
https://doi.org/10.1007/978-1-4842-9688-2

Vergadia, P. (2021, June). *Dataflow, the backbone of data analytics.* Retrieved from https://cloud.google.com/blog/topics/developers-practitioners/dataflow-backbone-data-analytics

Google Workspace (n.d.). *Nielsen: Collaborating across 100 countries for better consumer insights.* Retrieved from https://workspace.google.com/customers/nielsen.html

Index

A

AI Platform, 168
   capabilities, 51
   tools, 54–59
aiplatform.init method, 87
Apache Beam, 108, 156,
      180, 205
AutoML Tables, 57

B

Bar chart, 122, 126, 139
BigQuery, 35, 36, 121, 149, 162, 163,
      168, 196, 205
   analysis, 151
   and Cloud Storage, 11, 211
   dataset, 203, 210
   editor, 46, 51
   Google Cloud AI
      Platform, 33
   and Looker Studio, 134, 135
   ML model, 43–51
   Sandbox, 37–44, 135
   SQL queries, 36–43
   table, 155
   Vertex AI, 87–90
   web UI, 151

Bubble map, 141
Business intelligence, 36, 108, 121,
      122, 146, 147

C

Cloud AI Platform, 5, 7
   Google, 33, 35, 51–56, 100, 119
   Notebooks, 57
Cloud AutoML, 2, 57
Cloud-based platform, 2
Cloud CDN, 4
Cloud Data Fusion, 189, 204–211
Cloud ML Engine, 57
Cloud Pub/Sub, 149, 161
Cloud Storage, 211
Cloud Storage bucket, 66–69,
      154, 165–168
Code reviews, 195
Code search, 194
Colab, 11, 122
   data visualization, 142–146
   files, 45
   interface, 14, 15
   notebook, 42
   sheet, 43
Custom-trained model, 71–82

© Shitalkumar R. Sukhdeve and Sandika S. Sukhdeve 2023
S. R. Sukhdeve and S. S. Sukhdeve, *Google Cloud Platform for Data Science*,
https://doi.org/10.1007/978-1-4842-9688-2

Printed in the United States
by Baker & Taylor Publisher Services